Staged Cascades In Chemical Processing

PRENTICE-HALL INTERNATIONAL SERIES
IN THE PHYSICAL AND CHEMICAL ENGINEERING SCIENCES

AMUNDSON *Mathematical Methods in Chemical Engineering: Matrices and Their Application*
ARIS *Elementary Chemical Reactor Analysis*
ARIS *Introduction to the Analysis of Chemical Reactors*
ARIS *Vectors, Tensors, and the Basic Equations of Fluid Mechanics*
BALZHISER, SAMUELS, AND ELIASSEN *Chemical Engineering Thermodynamics*
BERAN AND PARRENT *Theory of Partial Coherence*
BOUDART *Kinetics of Chemical Processes*
BRIAN *Staged Cascades in Chemical Processing*
CROWE ET AL. *Chemical Plant Simulation*
DOUGLAS *Process Dynamics and Control, Volume 1, Analysis of Dynamic Systems*
DOUGLAS *Process Dynamics and Control, Volume 2, Control System Synthesis*
FREDRICKSON *Principles and Applications of Rheology*
HAPPEL AND BRENNER *Low Reynolds Number Hydrodynamics: With Special Applications to Particulate Media*
HIMMELBLAU *Basic Principles and Calculation in Chemical Engineering, 2nd Edition*
HOLLAND *Multicomponent Distillation*
HOLLAND *Unsteady State Processes with Applications in Multicomponent Distillation*
KOPPEL *Introduction to Control Theory with Applications to Process Control*
LEVICH *Physicochemical Hydrodynamics*
MEISSNER *Processes and Systems in Industrial Chemistry*
PERLMUTTER *Stability of Chemical Reactors*
PETERSEN *Chemical Reactor Analysis*
PRAUSNITZ *Molecular Thermodynamics of Fluid-Phase Equilibria*
PRAUSNITZ AND CHUEH *Computer Calculations for High-Pressure Vapor-Liquid Equilibria*
PRAUSNITZ, ECKERT, ORYE, AND O'CONNELL *Computer Calculations for Multicomponent Vapor-Liquid Equilibria*
WHITAKER *Introduction to Fluid Mechanics*
WILDE *Optimum Seeking Methods*
WILLIAMS *Polymer Science and Engineering*
WU AND OHMURA *The Theory of Scattering*

PRENTICE-HALL, INC.
PRENTICE-HALL INTERNATIONAL, INC., UNITED KINGDOM AND EIRE
PRENTICE-HALL OF CANADA, LTD., CANADA

Staged Cascades In Chemical Processing

P. L. THIBAUT BRIAN

Professor of Chemical Engineering
Massachusetts Institute of Technology

PRENTICE-HALL, INC.
Englewood Cliffs, N.J.

10 9 8 7 6 5 4 3 2 1

ISBN: 0–13–840280–9
Library of Congress Catalog Card Number: 79–181496

Printed in the United States of America

PRENTICE-HALL INTERNATIONAL, INC., *London*
PRENTICE-HALL OF AUSTRALIA, PTY. LTD., *Sydney*
PRENTICE-HALL OF JAPAN, INC., *Tokyo*
PRENTICE-HALL OF CANADA, LTD., *Toronto*
PRENTICE-HALL OF INDIA PRIVATE LTD., *New Delhi*

Contents

Preface

This text presents an introduction to staged cascades in chemical processing. The treatment assumes no knowledge of thermodynamics. Equilibrium between phases is described phenomenologically and is treated as experimentally determined, although the importance of the future study of thermodynamics is pointed out. Similarly, the idea of stage efficiency is introduced, but the application of fluid mechanics and mass transfer theory to the understanding of stage efficiency is left to future study.

With these simplifications, the description of the behavior of a single stage is quite straightforward involving only material balances, experimentally determined equilibrium relationships, and in some cases greatly simplified energy balances. This in turn permits focusing the discussion on *cascade theory*, the manner in which the individual stages interact to produce the behavior of the cascade as a whole.

This text has been developed for a first-term sophomore year course in chemical engineering, which has served simultaneously as an introduction to chemical engineering itself. The course could equally well be offered to freshmen.

Although a discussion of computer programming has not been included, digital computation should naturally accompany the application of this material; the iteration methods presented in Chapter 5 are obviously intended for use with a digital computer. Even when the cascade is described by closed-form solutions, as in Chapters 2 and 3, it is useful and practical to employ a computer in the solution of optimization problems in cascade design. Thus it is appropriate to visualize a "computation laboratory" accompanying the study of this material and to include study problems that require a digital computer for their solutions.

The incentive to offer this material as an introduction to chemical engineering in a university course comes from several sources. Paramount is the desire to expose a beginning student to some typical professional problems of chemical engineering. As engineering science courses rightly evolve in more rigorous and basic directions, it seems most undesirable to delay a student's contact with the professional flavor of process design, economic balance, and systems concepts until after he has travelled the long road of rigorous preparation in the basic and engineering sciences. This text is ideal for introducing these various professional engineering concepts to the student who has little preparation beyond a high school education. An early presentation, prior to or along with his beginning study of the engineering sciences, will add greatly to his perspective and motivation throughout his undergraduate years. He will approach the study of thermodynamics, fluid mechanics, and mass transfer with an awareness of at least some real engineering problems which require this knowledge for their elucidation.

Furthermore, this material is offered early in order to introduce automatic computation at the beginning of a course of study in chemical engineering. The practice and teaching of chemical engineering will employ computers in the future as naturally as slide rules were used in the past. Since the investment in learning computer programming is no greater for a freshman than for a senior, it is to the student's advantage to make it as early as possible so that he may utilize computers when and as they are helpful throughout his study of chemical engineering. But it is important to avoid an infatuation with computers for computers' sake alone; good judgment must always be exercised in choosing when and when not to call upon the aid of the computer. It is therefore most appropriate to introduce computation in the context of an engineering subject that really demands computation for its exploitation. The subject matter in this text is ideal for this purpose, emphasizing the subtle interactions of many stages whose individual behavior can be simply described with good accuracy.

This is by no means an exhaustive treatment of cascade design, but rather an introductory text. It attempts to clarify the important ideas of the subject by discussing them in depth from several points of view, but makes no pretense of covering all important methods of approaching these problems. It does, however, lay the foundation for self-study of the many topics which were not selected for inclusion.

Chapter 1 presents an introduction to separation processes and cascades in a chemical process, ideas which must be introduced here but which must be repeated throughout the text. Chapter 2 introduces the important concepts of staging and of countercurrent contacting in the simplest possible context, that of washing finely divided solids, in which a linear description is quite reasonable. The opportunity to couple the discussion with the description of a real chemical process is seized. Chapter 3 develops the linear model further in the context of liquid–liquid extraction, and exploits it to the fullest to explore the concept of the fractional cascade and the use of reflux.

Chapter 4 treats binary distillation in the classical manner but without the usual reliance upon thermodynamics, in order to emphasize the power of the McCabe–Thiele diagram as an aid to understanding. This material is especially well-suited for developing the ideas of economic balance in the design context. Finally, Chapter 5 presents multicomponent distillation problems, including iteration methods for the mathematical solution and also a discussion of multicomponent column performance.

The text contains numerous example problems which clarify the discussion and should be considered an integral part of the presentation. Unsolved problems are also included to serve as examples of the kinds of additional problem assignments that will contribute to an understanding of the material. In this context there is always the additional goal of giving the student a penetrating glimpse into the nature and character of industrial chemical processing.

P.L.T. BRIAN

Cambridge, Mass.

Introduction 1

Cascades of contacting stages are frequently employed in chemical processing, primarily in separating pure components from mixtures, and occasionally in accomplishing a chemical reaction. In each stage, two process streams are contacted and brought approximately to equilibrium with respect to each other. A number of such contacting stages are arranged in a cascade which produces the desired physical separation or chemical change. Equilibrium in the individual contacting stages is often closely approached, and thus an analysis of the behavior of a cascade of *equilibrium* stages yields much insight into the behavior of the real cascade and therefore plays an important role in the design of such systems.

1-1. Separation processes

Separating mixtures into essentially pure components is an important step in almost every chemical process, and it is not uncommon to find as much as half of the capital investment in a chemical plant required for carrying out such separations. The importance of separation processes in chemical processing is illustrated in Fig. 1-1, which is a sketch of a typical chemical process. The chemical reactor invariably requires a very pure feed stream because impurities in the feed would often result in undesirable side reactions, and also because in most catalytic processes even trace quantities

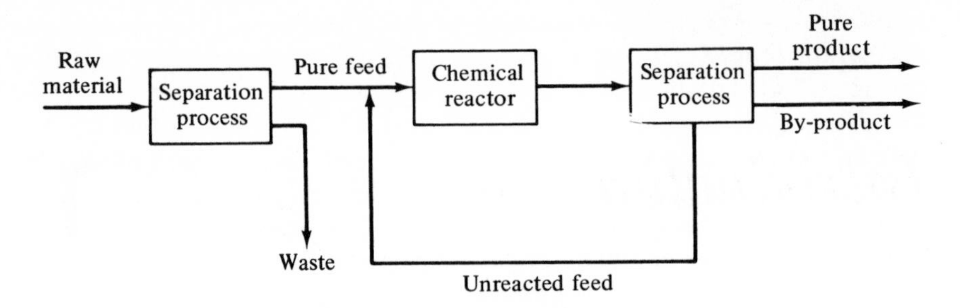

Fig. 1-1. Typical chemical process.

of certain impurities will poison the catalyst. But nature seldom provides the raw material in a suitably pure form; therefore, it is usual to find a separation process to purify the raw material before feeding it to the chemical reactor. The chemical reactor, on the other hand, while requiring a pure feed stream, almost never produces a pure product. Reversibility of the chemical reaction and/or a partial approach to chemical equilibrium in the reactor invariably result in unreacted feed material in the reactor effluent. In addition, by-products are often produced by the main reaction or by side reactions. Generally speaking, however, the desired product must be highly pure to be sold; therefore, another separation process is required to purify the reactor effluent into the pure product stream, the by-product streams, and the unreacted feed material, which will be recycled back to the reactor.

Because the cost of separating mixtures into pure components is an important part of the total cost in most chemical processes, it is not unusual to find that separation costs can often dictate the chemical route upon which the process will be based. Thus chemical process design often involves a fascinating interplay between the economics of various chemical reaction schemes and the economics of various separation processing schemes to arrive at an optimum process design.

1-2. Physical separation processes

The separation processes to be considered here are those involving mixtures of several components in gaseous or liquid solutions. Chemical techniques for separating pure components from such mixtures are occasionally employed, but such instances are not common, and the techniques employed are usually highly specific to the particular situation. By far the most common way of separating pure components from gaseous and liquid solutions is the use of physical methods of separation. In these processes, separation is usually achieved by bringing into contact two immiscible fluid phases in which the components to be separated distribute in different ratios because of differing solubility characteristics. Separation of the phases then results in some degree of separation of the components. For example, in distillation processes a vapor phase and a liquid phase are brought into contact. When equilibrium is achieved, the more volatile components are found to be distributed predominately in the vapor phase, while the less volatile components are distributed predominately in the liquid phase. The liquid and vapor phases can be separated by gravity, thus effecting a partial separation of the more volatile components from the less volatile components. Usually the volatilities of the components to be separated are not sufficiently different so as to enable the desired separation to be accomplished in a single contacting stage. Therefore, a number of stages are arranged together in a cascade which amplifies the separation achieved in each stage so as to produce the desired separation in the entire cascade. A distillation column is such a cascade employing countercurrent contacting of the vapor and the liquid phases in a series of contacting stages.

Often liquid-phase solutions contain components whose volatilities (at permissible temperatures) are so low that their distribution into the vapor phase would be so slight as to rule out distillation processes as an economic method of separation. In these cases, it is usual to seek a pair of immiscible liquids, such as an oil and water, into which the material to be separated will dissolve to an appreciable extent. When these two immiscible phases, containing the materials to be separated, are brought into contact, the components to be separated will distribute between the immiscible liquid phases. Generally speaking, the equilibrium ratio of two components in one liquid phase will be different from their ratio in the other phase, and the physical separation of the two liquid phases by gravity will therefore effect some degree of separation between the two components. As in distillation, a countercurrent cascade of many such contacting stages can be used to amplify the separation attained at each stage into the desired separation produced by the liquid–liquid extraction cascade.

1-3. Cascade design

The design of a cascade such as a distillation column or a liquid–liquid extraction cascade involves a consideration of the performance of each contacting stage and of the interactions of the various stages to produce the performance of the cascade as a whole. The performance of a single stage is usually expressed in terms of a stage efficiency, which represents the fractional approach to equilibrium achieved when the two streams are contacted in that stage. A detailed consideration of stage efficiency involves fluid mechanics and the rates of heat and mass transfer; this subject will not be considered here. Rather the focus here will be upon the performance of a cascade of equilibrium stages or of stages with assigned stage efficiencies, with the emphasis being upon the interactions among the various stages to produce the overall cascade performance. From a practical point of view such an approach is essentially complete for those situations in which the contacting stage efficiency is known to be almost 100 per cent. Even when the contacting efficiency is substantially less than 100 per cent, the division of the overall problem into the understanding of cascade performance on the one hand and the prediction of stage efficiency on the other hand is a very powerful method of bounding the problem and keeping it manageable.

As in most engineering design problems, it is important that reasonably accurate mathematical models of the process be formulated and solved in order that the performance of the proposed process may be calculated and its economic potential evaluated without actually building and operating the equipment. In this regard, the contacting stage of a physical separation process is usually susceptible to description by a relatively simple mathematical model. In general, equations are written that describe, for each stage in the cascade, the following:

1. Material balances for all components entering and leaving the stage.
2. An energy balance around the stage.
3. Relationships among concentrations, temperature, and pressure which represent equilibrium between the phases.

The science of thermodynamics is a powerful tool for understanding theoretically the relationships describing chemical equilibrium and phase equilibrium, but this subject will not be treated in detail here. Rather, the present treatment will consider the equilibrium relationships as having been determined experimentally, leaving the thermodynamic analysis of equilibrium to future study. In some of the processes to be considered in the present treatment, the energy balance around the stage is not crucial to a description of the performance of the stage, and thus it may be neglected. Even in distillation, where the energy balance is most certainly important, a detailed consideration of the energy balance can be avoided by making certain simplifying assumptions which, however, are actually quite accurate. Therefore, in the present treatment the mathematical model of a single stage will largely be obtained by writing material balances around the stage and by employing experimentally determined equilibrium relationships among the concentrations. This will be found to lead to a relatively simple description of each stage in the cascade. The major problem in mathematical modeling will appear in the form of the interactions of the various stages with each other, as represented by the simultaneous solution of the equations which describe the various stages.

The method of achieving a simultaneous solution of the equations describing the various stages will depend upon their form and complexity. For certain simple washing and extraction problems the equations will be linear, and closed-form algebraic expressions can be obtained by algebraic methods. When the equilibrium relationships are nonlinear, as in distillation, graphical methods of solving the equations can be used for the separation of binary mixtures. The graphical methods are especially useful as a method of visualizing and understanding the relationships among the various variables involved in the cascade design. For multicomponent separations the graphical techniques are not very useful, and if the equilibrium relationships are nonlinear, iterative techniques are employed for solving the system of simultaneous equations, usually with the aid of a digital computer.

2 Simple Linear Cascades: Washing of Finely Divided Solids

The basic ideas of staged cascades are perhaps best introduced in the context of a simple operation such as the washing of finely divided solids. The concept of equilibrium is especially simple in this case, and a linear mathematical model of the process is usually a good approximation. An important industrial example of this type of operation is the washing of bauxite mud in the Bayer alumina process.

2-1. Alumina mud washing

Aluminum metal is produced commercially by electrolytic reduction of alumina, Al_2O_3. Very high purity requirements for the finished metal product dictate that the alumina must be highly pure also. This high-purity alumina is produced by the Bayer process, which utilizes bauxite as the raw material. Bauxite is an ore which contains hydrated alumina together with a number of impurities, including iron oxide, titania, and silica. In the Bayer process the bauxite is reacted with an aqueous sodium hydroxide solution at high temperature and pressure. The alumina in the bauxite dissolves in the form of sodium aluminate, $NaAlO_2$, but the iron, silicon, and titanium compounds are insoluble and remain as a finely divided solid material suspended throughout the solution. This solid waste material, called *mud*, must be removed from the sodium aluminate solution and discarded.

First, the suspension of mud in sodium aluminate solution flows into a settling tank, as shown in Fig. 2-1. The feed, containing approximately 3 wt % solids suspended in the solution, is fed continuously to the middle

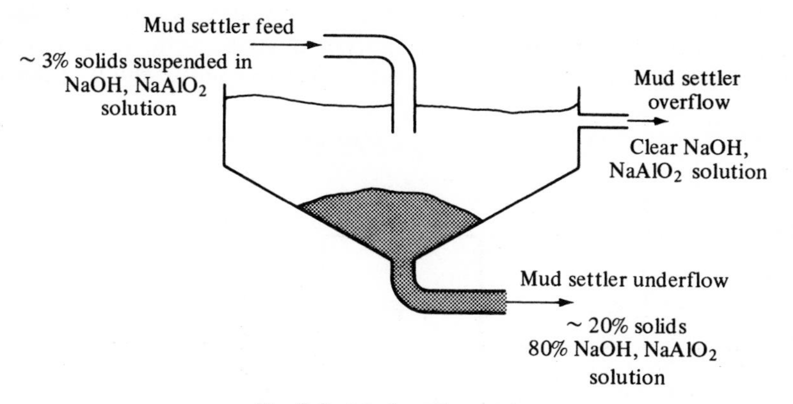

Fig. 2-1. Mud settling tank.

of the tank. The solid material settles to the bottom of the tank, where it is withdrawn as a mud phase, and the clear liquid is withdrawn from the top of the tank. The solid material is very finely divided; consequently, it settles, under the influence of gravity, at a very low settling velocity, typically of the order of 4 ft/hr. Therefore, the cross-sectional area of the tank must be sufficiently large so that the velocity of the liquid flowing up to the top of the tank will be low enough to avoid carrying much solid material overhead with the clear-liquor overflow stream. In practice the clear liquid withdrawn at the top contains very little solid material, but even this must be removed by a filtration step later in the process. In contrast, the mud stream flowing from the bottom of the tank is by no means free of sodium aluminate solution; typically the underflow stream will contain approximately 20 per cent solid material and 80 per cent liquid solution. This liquid content of the underflow stream depends upon the extent to which the small solid particles compact when they settle, and this is determined by the shape and size distribution of the solid particles, the design of the settling tank, and the fluid velocities in the tank. This mud phase is sufficiently fluid so that it can be pumped out of the bottom of the tank, but usually a rotating rake along the bottom of the tank is necessary to cause the mud to flow toward the center of the tank where the withdrawal pipe is located. The liquid solution is typically about 3.4 M in sodium ion content, approximately 60 per cent in the form of sodium aluminate and 40 per cent in the form of sodium hydroxide. This "soda" content of the liquid carried along with the mud stream is quite valuable, and most of it must be recovered by washing the mud before the solid material can be discarded.

SINGLE-STAGE WASHING

A single-stage system for washing the mud stream is shown in Fig. 2-2. The underflow stream from the mud settler is mixed with wash water and the mixture flows continuously to a second settling tank, which operates in

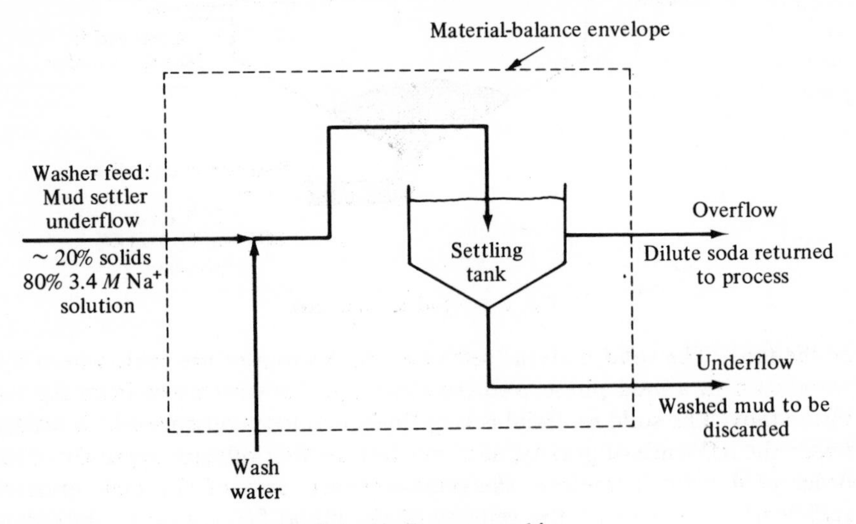

Fig. 2-2. Single-stage washing.

a fashion similar to the mud settler shown in Fig. 2-1. Here the solid material settles to the bottom of the tank and is withdrawn as an underflow stream, which is discarded. The overflow stream withdrawn from the top represents the soda recovered and returned to the process. Note, however, that this soda solution is dilute because of the wash water added, and this excess water returned to the process with the recovered soda must be evaporated at some other point in the Bayer alumina process. The soda recovery is by no means complete, because the underflow stream from the settling tank has a liquid content approximately equal to that of the underflow from the mud settler, typically of the order of 80 per cent liquid.

An equation relating the soda recovery to the quantity of wash water used will now be derived. Energy considerations are not important in this regard; therefore, an energy balance need not be written. Material balances for the solid material, the sodium salts, and the water are written around the system shown by the dashed envelope in Fig. 2-2. It is assumed that the system is operating continuously and at steady state. Steady-state operation is defined as operation in which the flow rate, solids content, and sodium-ion concentration of all entering and leaving streams are constant with respect to *time* (indeed these quantities are also independent of time at any point within the system). Therefore, the material-balance relationship for any component simplifies to the requirement that the rate of flow of this component into the system must equal its rate of flow from the system. The component flow rate is usually expressed as the product of the total flow rate and the component concentration. This may be the total *mass* flow rate multiplied by the component *mass fraction* or it may be the total *volumetric* flow rate multiplied by the *concentration* of the component. Various choices are available for the units in which the total flow rate and the component composition may be expressed, and the choice in any situation will usually be based upon convenience, or more particularly upon which type of flow is more nearly constant in a given situation.

In the present case, the density of the aqueous soda solution will vary with the soda concentration in the solution, and it seems more appropriate to assume a constant value for the volume per cent liquid in the underflow stream than to assume a constant value for the weight per cent liquid. Likewise, it is a good approximation to assume that when a strong soda solution is mixed with water to form a dilute soda solution, the volume of the dilute solution is essentially equal to the sum of the volumes of the strong solution and of the water. With these two simplifying assumptions, the material balances for the present case can be written without considering the manner in which the density of the soda solution varies with its concentration, and the soda material balance is appropriately written in terms of volumetric liquid flow rates multiplied by sodium-ion concentrations in moles per liter.

It is convenient to assume that the solids carryover in the overflow stream from the settling tank is negligible, and this would be a good appxoximation to industrial practice. With this simplification the material balance on the solid material simply requires that the mass flow rate of the solids in the underflow stream from the settling tank be equal to the mass flow rate of solids in the feed stream to the washing system. This solids mass flow rate will be denoted by the symbol S. It is likewise convenient to assume that the volume of liquid per unit mass of solid, denoted by the symbol L, has the same value in the washer underflow stream as it does in the washer feed stream. It follows then that SL, the volumetric flow rate of liquid carried with the solids, is the same in the washer feed and in the washer underflow streams. From this it follows that the volumetric flow rate of the washer overflow stream is equal to the volumetric flow rate of the wash-water stream, which is denoted by the symbol W, because of the assumption of the additivity of liquid volumes upon mixing strong soda solutions with water. Note that these assumptions eliminate the need for stating a water material balance formally and replace the water material balance by a *liquid volume balance.*

Assuming that the wash water is essentially free of sodium ions, the soda balance can be expressed as

$$SLC_F = SLC_U + WC_O \tag{2-1}$$

where C_F, C_U, and C_O represent the sodium-ion concentrations in the aqueous liquid phase in the feed, underflow, and overflow streams, respectively. These concentrations are expressed as moles per unit of solution volume. The concept of equilibrium in this type of contacting operation is simply that the mud feed stream and the wash-water stream are mixed sufficiently well that the aqueous solution entering the settling tank is of uniform composition throughout and, therefore, that the soda concentration in the aqueous phase in the underflow stream is the same as that in the overflow stream. Therefore, the equilibrium expression is particularly simple in this case:

$$C_O = C_U \tag{2-2}$$

Substitution of Eq. (2-2) into Eq. (2-1) and rearrangement yields

$$\frac{C_U}{C_F} = \frac{1}{1 + W/SL} \tag{2-3}$$

Since the volumetric flow rate of liquid in the underflow stream is equal to that in the feed stream, the left-hand side of Eq. (2-3) is equal to the fraction of the soda in the washer feed lost with the discarded mud stream. The ratio W/SL is simply the volume of wash water added per unit volume of liquid phase flowing in the underflow stream. This ratio will be given the symbol E and will be called an *extraction factor* because it represents a measure of how well the washing operation can extract the soda from the mud stream before the mud is discarded. Denoting the fraction of the soda in the washer feed recovered in the washer operation by the symbol R, Eq. (2-3) can be rewritten as

$$1 - R = \frac{1}{1 + E} \tag{2-4}$$

EXAMPLE PROBLEM 2-1 The mud-settler underflow stream in a proposed Bayer alumina plant will contain waste solids flowing at a rate of 1000 tons of solid material per day and will have a liquid content of 4 ml of aqueous soda solution per gram of solid material. If a single-stage mud-washing system is employed, at what rate must wash water be fed to the system in order to recover 90 per cent of the soda from the mud stream? Repeat the calculation for 95 per cent recovery and for 98 per cent recovery. Assume that the liquid content of the washer underflow is also 4 ml of aqueous solution per gram of solid material.

Solution: For 90 per cent of soda recovery, $R = 0.9$, and Eq. (2-4) requires that

$$E \equiv \frac{W}{SL} = 9$$

Since $L = 4$ ml of solution per gram of solid material,

$$\frac{W}{S} = \left(\frac{9 \text{ ml of wash water}}{\text{ml of solution}}\right)\left(\frac{4 \text{ ml of solution}}{\text{g of solid}}\right) = \frac{36 \text{ ml of wash water}}{\text{g of solid}}$$

But, since the density of the wash water is essentially 1 g/ml,

$$\frac{\rho_W W}{S} = \frac{36 \text{ g of wash water}}{\text{g of solid material}} = \frac{36 \text{ tons of wash water}}{\text{ton of solid}}$$

Therefore, the required wash-water rate is

$$\rho_W W = \left(\frac{36 \text{ tons of wash water}}{\text{tons of solid}}\right)\left(\frac{1000 \text{ tons of solid}}{\text{day}}\right)$$

$$= 36{,}000 \text{ tons of wash water/day}$$

$$= \frac{(36{,}000 \text{ tons/day})(2000 \text{ lb/ton})}{(1440 \text{ min/day})} = 50{,}000 \text{ lb/min}$$

Since the density of water is also 8.34 lb/gal, the wash-water rate can be expressed as

$$W = \frac{50{,}000 \text{ lb/min}}{8.34 \text{ lb/gal}} = 6000 \text{ gal/min}$$

The above calculation employed a knowledge of the density of water in grams per milliliter and in pounds per gallon. A more straightforward approach can be taken knowing the conversion factors from pounds to grams, pounds to tons, and gallons to milliliters:

$$W = 9SL = \frac{(9)(1000 \text{ tons/day})(4 \text{ ml/g})(453.6 \text{ g/lb})(2000 \text{ lb/ton})}{(1440 \text{ min/day})(3785 \text{ ml/gal})} = 6000 \text{ gal/min}$$

For 95 per cent soda recovery, $R = 0.95$, and Eq. (2-4) requires that $E = 19$. The required wash-water rate is

$$W = \left(\frac{19}{9}\right)\left(\frac{6000 \text{ gal}}{\text{min}}\right) = 12{,}700 \text{ gal/min}$$

For 98 per cent soda recovery, $E = 49$ and $W = 32{,}700$ gal/min

EXAMPLE PROBLEM 2-2 For the three cases considered in Example Problem 2-1, calculate the diameter of the washer settling tank, assuming that only one tank is employed for settling the entire stream. As an approximation, assume that the cross-sectional area of the tank must be large enough so that the average velocity of the liquid flowing up to the top of the tank to the overflow pipe should not exceed 3 ft/hr. This average liquid velocity is defined as the volumetric flow rate of liquid in the overflow stream divided by the cross-sectional area of the tank.

Solution: For 90 per cent soda recovery, the wash-water rate is 6000 gal/min, as given in Example Problem 2-1. Therefore, the dilute-soda overflow stream from the washer settling tank is also 6000 gal/min. The required cross-sectional area of the settling tank is

$$A = \frac{(6000 \text{ gal/min})(60 \text{ min/hr})}{(3 \text{ ft/hr})(7.48 \text{ gal/ft}^3)} = 16{,}000 \text{ ft}^2$$

and the required diameter is

$$d = \sqrt{\frac{4}{\pi}(16{,}000\ \text{ft}^2)} = 143\ \text{ft}$$

For 95 per cent soda recovery, the required tank diameter is

$$d = (143\ \text{ft})\sqrt{\frac{19}{9}} = 208\ \text{ft}$$

For 98 per cent soda recovery, the required tank diameter is 334 ft.

MULTISTAGE CROSS-FLOW CASCADE

In multistage cross-flow washing, the mud is washed a number of times, each step employing only a fraction of the total wash water. Figure 2-3 shows a diagram of a four-stage cross-flow washing system with a uniform

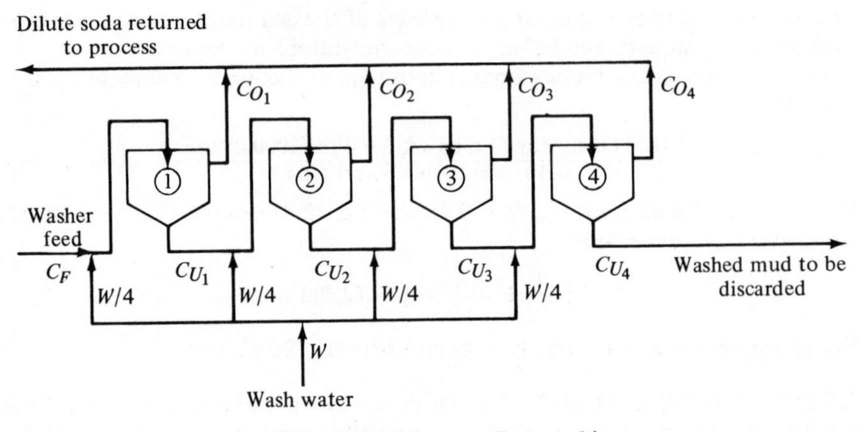

Fig. 2-3. Four-stage cross-flow washing.

distribution of the wash water among the various stages. Thus, one fourth of the total wash-water flow is employed in each of the four washing steps. The feed to any stage in the cascade consists of a mixture of fresh water and the underflow stream from the previous stage. The underflow stream from the final stage represents the washed mud to be discarded, while the overflow streams from all the stages are mixed together to form the recovered soda stream, which is returned to the process. The extension of this scheme to any number of stages is quite straightforward.

Referring to Fig. 2-2 and the derivation of Eq. (2-3), it is apparent that an analogous derivation for the first washing step in Fig. 2-3 will lead to the equation

$$\frac{C_{U_1}}{C_F} = \frac{1}{1 + (W/4SL)} \tag{2-5}$$

For the second, third, and fourth washing steps,

$$\frac{C_{U_2}}{C_{U_1}} = \frac{1}{1 + (W/4SL)} \tag{2-6}$$

$$\frac{C_{U_3}}{C_{U_2}} = \frac{1}{1 + (W/4SL)} \tag{2-7}$$

$$\frac{C_{U_4}}{C_{U_3}} = \frac{1}{1 + (W/4SL)} \tag{2-8}$$

Multiplying Eq. (2-5), (2-6), (2-7), and (2-8) together yields

$$\frac{C_{U_4}}{C_F} = \frac{1}{(1 + [W/4SL])^4} \tag{2-9}$$

In this derivation it has been assumed that the feed stream and the underflow streams from all the washing steps have the same liquid content, L volumes of liquid per unit mass of solid. Therefore, the left-hand side of Eq. (2-9) represents the fraction of the soda in the feed stream that is discarded. Equation (2-9) can therefore be rewritten as

$$1 - R = \frac{1}{(1 + E/4)^4} \tag{2-10}$$

An analogous derivation for a cross-flow washing system with N stages with equal distribution of wash water to all stages yields the result

$$1 - R = \frac{1}{(1 + E/N)^N} \qquad (\text{for } N = 1, 2, 3, \ldots) \tag{2-11}$$

where E is defined as the total wash water fed to the entire system divided by the liquid flow rate in the feed stream. As the number of stages tends to infinity, Eq. (2-11) approaches the form

$$1 - R \longrightarrow e^{-E} \qquad (\text{for } N \longrightarrow \infty) \tag{2-12}$$

which follows directly from the definition of the exponential function.

EXAMPLE PROBLEM 2-3 Referring to Example Problem 2-1, what would be the required wash-water rate for 95 per cent soda recovery if cross-flow washing were employed with four stages and with equal distribution of the wash water among the stages? Repeat the calculation for 10 stages, 40 stages, and an infinite number of stages.

Solution: For 95 per cent soda recovery, R equals 0.95, and with $N = 4$, Eq. (2-11) requires that $E = 4.47$. This is to be compared with $E = 19$ for 95 per cent recovery with a single stage as given in Example Problem 2-1. The corresponding wash-water flow rate from Example Problem 2-1 was 12,700 gal/min; therefore, the required wash-water rate with four-stage washing is that rate multiplied by the ratio

4.47/19, or 2990 gal/min. The results for other values of N, the number of stages in the cascade, all for 95 per cent recovery, are given in Table 2-1. The values tabulated illustrate the large reduction in the wash-water requirement that can be achieved by using a multistage cross-flow cascade. It is also apparent that (for 95 per cent recovery) the percentage reduction in the wash-water requirement achieved by going from a single-stage to a two-stage system is greater than the additional reduction achieved by adding an infinite number of stages.

Table 2-1

N	E	W, gal/min
1	19	12,700
2	6.94	4,630
4	4.47	2,990
10	3.50	2,340
40	3.11	2,080
∞	2.995	2,000

2-2. Optimum allocation of wash water

The preceding analysis of a multistage cross-flow cascade was based upon the assumption that the wash water would be divided equally among the various washing stages. If this is not the case, equations of the form of Eq. (2-5), (2-6), (2-7), and (2-8) would be modified to contain the actual wash-water flow rate fed to each individual stage, and the final result, Eq. (2-11), would take the form

$$\frac{1}{1-R} = (1 + Ef_1)(1 + Ef_2)(1 + Ef_3) \cdots (1 + Ef_N) \qquad (2\text{-}13)$$

where f_j represents the fraction of the total wash water fed to the jth stage. It follows from this definition that

$$f_1 + f_2 + f_3 + \cdots + f_N = 1 \qquad (2\text{-}14)$$

It can be shown that for a given number of stages and for a given total wash-water rate (that is, a given value of E) the recovery is maximized when the wash water is divided equally among the various stages. Consider a four-stage cross-flow cascade. With the definition

$$\phi \equiv \frac{1}{1-R} \qquad (2\text{-}15)$$

Eqs. (2-13) and (2-14) become

$$\phi = (1 + Ef_1)(1 + Ef_2)(1 + Ef_3)(1 + Ef_4) \qquad (2\text{-}16)$$

$$f_1 + f_2 + f_3 + f_4 = 1 \qquad (2\text{-}17)$$

Eliminating f_4 between Eqs. (2-16) and (2-17) yields

$$\phi = (1 + Ef_1)(1 + Ef_2)(1 + Ef_3)[1 + E(1 - f_1 - f_2 - f_3)] \tag{2-18}$$

The conditions that will maximize ϕ are given by

$$\frac{\partial \phi}{\partial f_1} = (1 + Ef_2)(1 + Ef_3)[1 + E(1 - f_1 - f_2 - f_3)]E - (1 + Ef_1)(1 + Ef_2)(1 + Ef_3)E = 0 \tag{2-19}$$

$$\frac{\partial \phi}{\partial f_2} = (1 + Ef_1)(1 + Ef_3)[1 + E(1 - f_1 - f_2 - f_3)]E - (1 + Ef_1)(1 + Ef_2)(1 + Ef_3)E = 0 \tag{2-20}$$

$$\frac{\partial \phi}{\partial f_3} = (1 + Ef_1)(1 + Ef_2)[1 + E(1 - f_1 - f_2 - f_3)]E - (1 + Ef_1)(1 + Ef_2)(1 + Ef_3)E = 0 \tag{2-21}$$

Employing Eq. (2-17), these three equations can be reduced to

$$1 + Ef_4 = 1 + Ef_1 \tag{2-22}$$

$$1 + Ef_4 = 1 + Ef_2 \tag{2-23}$$

$$1 + Ef_4 = 1 + Ef_3 \tag{2-24}$$

from which it follows that

$$f_1 = f_2 = f_3 = f_4 = \tfrac{1}{4} \tag{2-25}$$

This is the condition for achieving a maximum in ϕ, which can be verified by noting that the second partial derivatives of ϕ with respect to f_1, f_2, and f_3 are negative. For example, differentiating Eq. (2-19) with respect to f_1 yields

$$\begin{aligned}\frac{\partial^2 \phi}{\partial f_1^2} &= -(1 + Ef_2)(1 + Ef_3)E^2 - (1 + Ef_2)(1 + Ef_3)E^2 \\ &= -2(1 + Ef_2)(1 + Ef_3)E^2\end{aligned} \tag{2-26}$$

from which it is obvious that the second partial derivative is negative. This is also true for the second partial derivatives with respect to f_2 and f_3. Obviously, maximizing ϕ will minimize the fraction of the soda not recovered and will, therefore, maximize the soda recovery. The extension of this analysis to any number of stages is quite straightforward, and it is obvious that the recovery will be maximized when the wash water is divided uniformly among the various stages.

If the liquid content, L, is not the same for all the mud streams fed to the various washing stages, then the maximum recovery will not occur when the wash water is divided equally among the various stages but rather when the wash water is allocated such that the sum of the liquid rates fed as fresh wash water and as water accompanying the mud stream is the same for all stages. For example, consider a four-stage cross-flow washing system, and let L_j be the liquid content of the mud stream *fed to* the jth stage. The material-balance equations, analogous to Eqs. (2-5), (2-6), (2-7), and (2-8), become

$$\frac{C_{U_1}}{C_F} = \frac{1}{1 + (Wf_1/SL_1)} \tag{2-27}$$

$$\frac{C_{U_2}}{C_{U_1}} = \frac{1}{1 + (Wf_2/SL_2)} \tag{2-28}$$

$$\frac{C_{U_3}}{C_{U_2}} = \frac{1}{1 + (Wf_3/SL_3)} \tag{2-29}$$

$$\frac{C_{U_4}}{C_{U_3}} = \frac{1}{1 + (Wf_4/SL_4)} \tag{2-30}$$

The fraction of the soda unrecovered is

$$\begin{aligned} 1 - R &= \frac{C_{U_4}L_5}{C_F L_1} \\ &= \frac{(L_5/L_1)}{(1 + [Wf_1/SL_1])(1 + [Wf_2/SL_2])(1 + [Wf_3/SL_3])(1 + [Wf_4/SL_4])} \end{aligned} \tag{2-31}$$

In this nomenclature L_1 and L_5 represent the liquid contents of the washer feed and the washed mud, respectively. Equations (2-15), (2-17), and (2-31) can be combined to give

$$\phi = \frac{L_1}{L_5}\left(1 + \frac{Wf_1}{SL_1}\right)\left(1 + \frac{Wf_2}{SL_2}\right)\left(1 + \frac{Wf_3}{SL_3}\right)\left[1 + \frac{W}{SL_4}(1 - f_1 - f_2 - f_3)\right] \tag{2-32}$$

Setting the derivative with respect to f_1 equal to zero yields

$$\begin{aligned} \frac{\partial \phi}{\partial f_1} = \frac{L_1}{L_5}\left(1 + \frac{Wf_2}{SL_2}\right)\left(1 + \frac{Wf_3}{SL_3}\right)\left[1 + \frac{W}{SL_4}(1 - f_1 - f_2 - f_3)\right]\frac{W}{SL_1} \\ - \frac{L_1}{L_5}\left(1 + \frac{Wf_1}{SL_1}\right)\left(1 + \frac{Wf_2}{SL_2}\right)\left(1 + \frac{Wf_3}{SL_3}\right)\frac{W}{SL_4} = 0 \end{aligned} \tag{2-33}$$

This can be simplified to

$$\frac{1 + (Wf_4/SL_4)}{L_1} = \frac{1 + (Wf_1/SL_1)}{L_4} \tag{2-34}$$

and still further to

$$SL_4 + Wf_4 = SL_1 + Wf_1 \tag{2-35}$$

The terms on the left-hand side of Eq. (2-35) represent the rate of flow of liquid into the fourth stage in the mud stream (which is the underflow from the third stage) and in the wash water allocated to the fourth stage. Similarly, the right-hand side of Eq. (2-35) represents the total liquid flow rate into the first stage, in the mud feed and in the wash water allocated to the first stage.

In a similar manner, partially differentiating Eq. (2-32) with respect to f_2 and f_3 and equating these derivatives to zero results in

$$SL_4 + Wf_4 = SL_2 + Wf_2 \tag{2-36}$$

$$SL_4 + Wf_4 = SL_3 + Wf_3 \tag{2-37}$$

It is apparent that these equations require that the total liquid fed to a given stage, representing the sum of the wash-water allocation to that stage and the liquid content of the solids feed to that stage, should be the same for all stages if the recovery is to be maximized. The extension of this analysis to any number of stages is quite straightforward, and it is apparent that this generalization applies irrespective of the number of stages. Equation (2-31) can be rewritten as

$$1 - R = \frac{S^4 L_2 L_3 L_4 L_5}{(SL_1 + Wf_1)(SL_2 + Wf_2)(SL_3 + Wf_3)(SL_4 + Wf_4)} \tag{2-38}$$

If the wash water is allocated so that the total liquid flow to each stage is the same, it is apparent from Eqs. (2-35), (2-36), and (2-37) that the four factors in the denominator on the right-hand side of Eq. (2-38) are all equal; thus, Eq. (2-38) can be rewritten as

$$1 - R = \frac{(L_2 L_3 L_4 L_5)/L_1^4}{(1 + Wf_1/SL_1)^4} \tag{2-39}$$

For N stages the analogous equation is

$$1 - R = \frac{(L_2 L_3 L_4 \ldots L_{N+1})/L_1^N}{(1 + Wf_1/SL_1)^N} \tag{2-40}$$

Since keeping the total liquid flow the same for all stages will maximize the recovery for any fixed number of stages and fixed total wash-water rate, it will also minimize the required wash-water rate for a fixed recovery and a fixed number of stages. But it does not necessarily follow that this represents the economic optimum allocation of the wash water, because the cost of wash water and the cost of soda lost with the mud will generally not be the only important costs to be considered in the economic balance. The capital investment in the settling tanks and other equipment and the cost of an automatic control system and/or the operating labor to control the process must also be considered in the economic balance; these considerations might dictate that the economically optimum way in which to allocate the wash water will not correspond to maintaining the same total liquid flow rate into all the washing stages.

EXAMPLE PROBLEM 2-4 Referring to Example Problems 2-1 and 2-3, consider a four-stage cross-flow washing system. Assume that the liquid content, L, is equal to 3.0, 3.5, 4.0, 4.5, and 5.0 ml of liquid per gram of solid in the washer-feed stream and in the underflow streams *leaving* the first, second, third, and fourth washing stages, respectively. Determine the minimum total wash rate that will achieve a 95 per cent soda recovery, and for this condition determine how the wash water should be allocated to the various stages. If, instead, this same total amount of wash water were divided uniformly among the stages, what soda recovery would result?

Solution: In the nomenclature of Eq. (2-40), the liquid contents assumed in this problem are

$$L_1 = 3.0, \quad L_2 = 3.5, \quad L_3 = 4.0, \quad L_4 = 4.5, \quad L_5 = 5.0$$

Substituting these values plus $R = 0.95$ and $N = 4$ into Eq. (2-40) yields

$$\left(1 + \frac{Wf_1}{SL_1}\right)^4 = (20)\frac{(3.5)(4.0)(4.5)(5.0)}{(3.0)^4}$$

Thus

$$\frac{Wf_1}{SL_1} = 1.97$$

or

$$Wf_1 + SL_1 = 2.97SL_1$$

But from Eqs. (2-35) and (2-36) it follows that

$$Wf_2 + SL_2 = Wf_1 + SL_1 = 2.97\,SL_1$$

Therefore,

$$\frac{Wf_2}{SL_1} = 2.97 - \frac{L_2}{L_1} = 2.97 - \frac{3.5}{3.0} = 1.80$$

Similarly,

$$\frac{Wf_3}{SL_1} = 2.97 - \frac{L_3}{L_1} = 2.97 - \frac{4.0}{3.0} = 1.64$$

$$\frac{Wf_4}{SL_1} = 2.97 - \frac{L_4}{L_1} = 2.97 - \frac{4.5}{3.0} = 1.47$$

Summing these four values yields

$$\frac{Wf_1}{SL_1} + \frac{Wf_2}{SL_1} + \frac{Wf_3}{SL_1} + \frac{Wf_4}{SL_1} = \frac{W}{SL_1} = 1.97 + 1.80 + 1.64 + 1.47 = 6.88$$

The total wash-water requirement is

$$W = \frac{(1000 \text{ tons/day})(3 \text{ ml/g})(6.88)(2000 \text{ lb/ton})(453.6 \text{ g/lb})}{(1440 \text{ min/day})(3785 \text{ ml/gal})} = 3440 \text{ gal/min}$$

and the fraction of this total that should be allocated to each stage is

$$f_1 = \frac{1.97}{6.88} = 0.288$$

$$f_2 = \frac{1.80}{6.88} = 0.261$$

$$f_3 = \frac{1.64}{6.88} = 0.238$$

$$f_4 = \frac{1.47}{6.88} = 0.213$$

If, instead of these allocations, the wash water is divided uniformly among the stages, the value of f is 0.25 for all stages, and

$$\frac{Wf_1}{SL_1} = (6.88)(0.25) = 1.72$$

$$\frac{Wf_2}{SL_2} = (1.72)\left(\frac{3.0}{3.5}\right) = 1.47$$

$$\frac{Wf_3}{SL_3} = (1.72)\left(\frac{3.0}{4.0}\right) = 1.29$$

$$\frac{Wf_4}{SL_4} = (1.72)\left(\frac{3.0}{4.5}\right) = 1.15$$

Substitution of these values into Eq. (2-31) yields

$$1 - R = \frac{5.0/3.0}{(2.72)(2.47)(2.29)(2.15)} = 0.0504$$

and thus the recovery drops from 95 to 94.96 per cent. In this case, it is seen that the recovery is not very sensitive to the way in which the wash water is allocated, and the simplicity of uniform allocation might dictate that this type of operation be used.

MULTISTAGE COUNTERCURRENT CASCADE

In the single-stage washing system shown in Fig. 2-2, the soda concentration in the dilute-soda stream returned to the process is equal to the soda concentration in the liquid carried with the discarded mud stream. Increasing the rate of addition of wash water will improve soda recovery, but the recovered soda will be more dilute in the stream returned to the process. In contrast to this, in the four-stage cross-flow system shown in Fig. 2-3, the recovered soda is returned to the process at a concentration substantially higher than the soda concentration in the liquid carried with the discarded mud stream. The soda concentration in the overflow from the fourth stage is equal to the soda concentration in the liquid carried with the discarded mud, but the overflow streams from the first, second, and third stages are more concentrated. Since the soda stream returned to the process is a mixture of these overflow streams, its soda concentration is higher than that in the washed mud, and the wash-water requirement for a given recovery is lower than that for a single-stage system.

Consider, for example, a four-stage cross-flow system with $E = 4.47$. According to Eq. (2-10), this corresponds to 95 per cent recovery; therefore, the soda concentration in the fourth-stage overflow and underflow streams is only $\frac{1}{20}$ of the soda concentration in the washer feed. According to Eqs. (2-5), (2-6), (2-7), and (2-8), the soda concentrations, relative to the soda concentration in the washer feed, are equal to 0.472, 0.223, 0.1053, and

or finally,

$$\frac{C_{U_1}}{C_F} = \frac{1}{1 + E + E^2 + E^3 + E^4} \tag{2-54}$$

Since the liquid contents of all mud streams are assumed to be equal, the left-hand side of Eq. (2-54) is equal to the fraction of the soda not recovered. The straightforward generalization for any number of stages leads to

$$1 - R = \frac{1}{1 + E + E^2 + E^3 + \cdots + E^N} \tag{2-55}$$

where N is the number of stages in the countercurrent cascade.

Equation (2-55) can be written in a more convenient form by using the identity

$$1 + E + E^2 + E^3 + \cdots + E^N \equiv \frac{1 - E^{N+1}}{1 - E} \tag{2-56}$$

The validity of this identity can be demonstrated by carrying out, with formal long division, the division indicated by the ratio on the right-hand side of Eq. (2-56), or it can be demonstrated by the following procedure. First adopt the definition

$$\lambda \equiv 1 + E + E^2 + E^3 + \cdots + E^N \tag{2-57}$$

Subtracting 1 from both sides of Eq. (2-57) and then dividing by E gives

$$\frac{\lambda - 1}{E} = 1 + E + E^2 + \cdots E^{N-1} \tag{2-58}$$

Subtracting Eq. (2-58) from Eq. (2-57) yields

$$\lambda - \frac{\lambda - 1}{E} = E^N \tag{2-59}$$

which can be rearranged to

$$\lambda = \frac{1 - E^{N+1}}{1 - E} \tag{2-60}$$

Equations (2-57) and (2-60) are equivalent to Eq. (2-56).

Using Eq. (2-56), Eq. (2-55) can be simplified to the form

$$1 - R = \frac{1 - E}{1 - E^{N+1}} \tag{2-61}$$

This equation expresses the recovery obtained in a linear countercurrent cascade in terms of the number of stages in the cascade and the extraction factor, E. It was first presented by Kremser [1] for the design of gas-absorption columns. For the special case $E = 1$, the right-hand side of Eq. (2-61) can be evaluated with the help of L'Hospital's rule to give

$$1 - R = \frac{1}{N + 1} \qquad (\text{for } E = 1) \tag{2-62}$$

It is apparent from Eq. (2-62) that by using a sufficiently large number of stages in the countercurrent cascade, any desired recovery can be achieved with $E = 1$, which corresponds to a wash-water addition rate equal to the volumetric feed rate of liquid with the washer feed mud stream. For example, a 99 per cent recovery can be obtained by using 99 stages and a wash-water rate corresponding to $E = 1$. In contrast, Eq. (2-12) reveals that 4.6 times as much wash water would be required in a cross-flow cascade to achieve 99 per cent recovery even with an infinite number of stages.

For a countercurrent cascade with a number of stages approaching infinity, when E is less than unity, Eq. (2-61) simplifies to

$$R \longrightarrow E \qquad (\text{for } N \longrightarrow \infty, E < 1) \tag{2-63}$$

An overall soda material balance for the entire cascade can be written as

$$WC_{ON} = RSLC_F \tag{2-64}$$

The left-hand side of Eq. (2-64) represents the soda content of the overflow from the Nth stage, which is the stream which returns the recovered soda to the process. The right-hand side of Eq. (2-64) is simply the fractional recovery multiplied by the soda content of the washer feed stream. Equation (2-64) can be rearranged to

$$\frac{C_{ON}}{C_F} = \frac{R}{E} \tag{2-65}$$

which shows that the factor R/E represents the dilution of the recovered soda relative to the liquid carried with the washer feed stream. It is apparent from Eqs. (2-63) and (2-65) that as the number of stages in a countercurrent cascade approaches infinity, the soda concentration in the recovered soda stream approaches that in the liquid carried along with the washer feed stream, provided that E is less than unity. This obviously represents the use of the minimum possible wash water to accomplish a given soda recovery. It should be noted that if a wash-water rate corresponding to a value of E greater than unity is employed, R will approach unity as N approaches infinity, as indicated by Eq. (2-61), but the soda concentration in the overflow stream from the Nth stage will not approach C_F.

EXAMPLE PROBLEM 2-5 Repeat Example Problem 2-3 for a countercurrent washing cascade.

Solution: With four stages and $R = 0.95$, Eq. (2-61) becomes

$$0.05 = \frac{1 - E}{1 - E^5}$$

The value of E must be greater than unity because it is apparent from Eq. (2-62) that if $E = 1$, the recovery will only be 80 per cent with a four-stage cascade. Therefore, the above equation is solved by trial and error for E, with the trial procedure starting with values of E somewhat greater than unity. The result is

$$E = 1.734$$

which corresponds to a wash-water rate of 1160 gal/min, obtained by simple ratio from the results of Example Problem 2-1. Table 2-2 gives results for other values of

Table 2-2

N	E	W, *gal/min*
2	3.89	2600
4	1.734	1160
10	1.147	766
40	0.96	641
∞	0.95	635

N, the number of stages in the countercurrent cascade, all for 95 per cent recovery. When the values tabulated are compared with the results of Example Problem 2-3, it is seen that for a given number of stages the wash-water requirement of the countercurrent cascade is lower than that for the cross-flow cascade by a factor of 2 to 3. This is for a recovery of 95 per cent. At much higher recoveries, the countercurrent cascade would produce even greater savings in wash water. For example, with R equal to 0.9999 and N approaching infinity, the value of E required for a cross-flow cascade is computed from Eq. (2-12) to be 9.21, as compared with a value of 0.9999 for a countercurrent cascade. Thus in this case the cross-flow system requires more than nine times as much wash water as the countercurrent system.

2-3. Generality of the Kremser equation

The Kremser equation describes the behavior of any simple countercurrent cascade in which the material balance and equilibrium relationships are linear. Thus it finds use not only in the design of mud-washing systems but also in the design of other processes, such as gas absorption and liquid–liquid extraction. In such cases, the extraction factor, E, is generalized to include the slope of the equilibrium line, as discussed in Chapter 3. Figure 2-5 shows two graphical representations of Eq. (2-61), as presented by

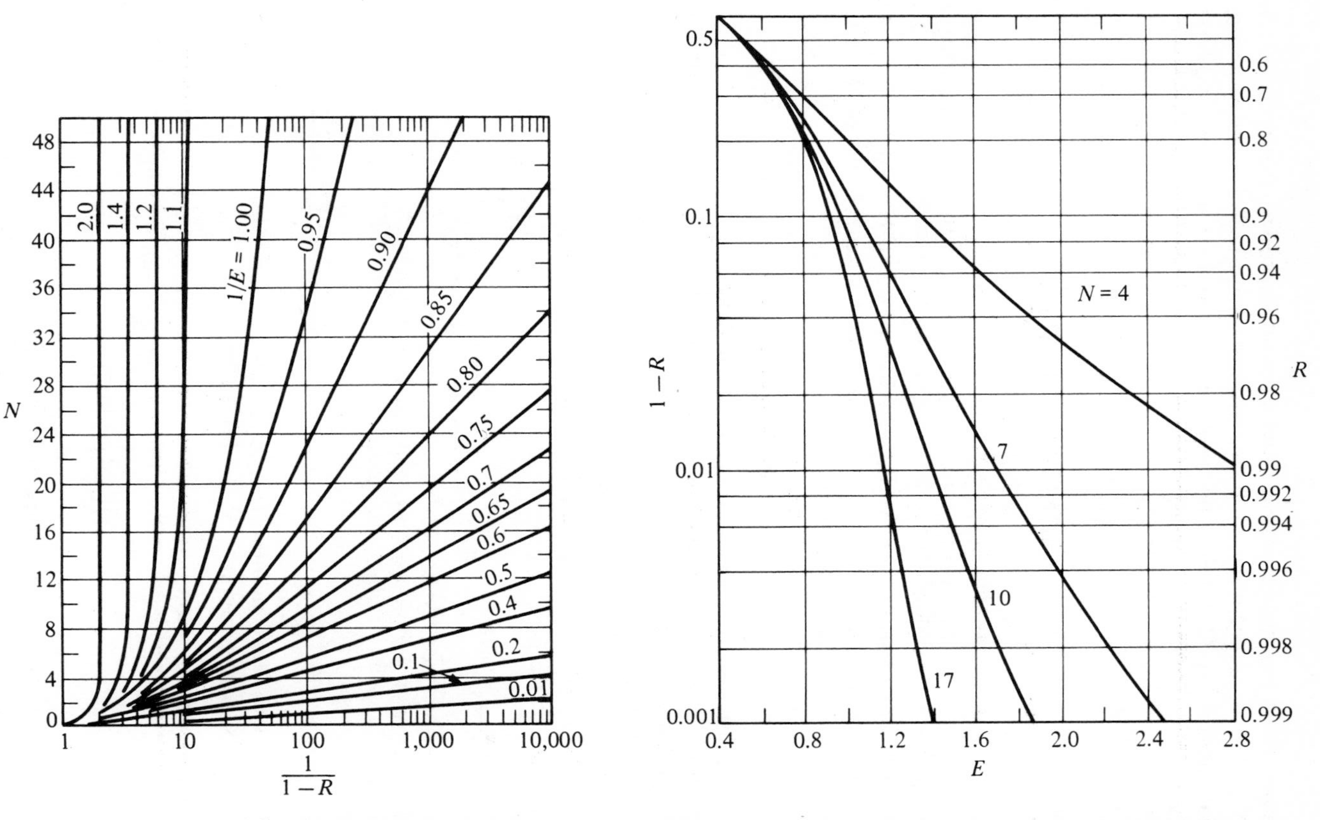

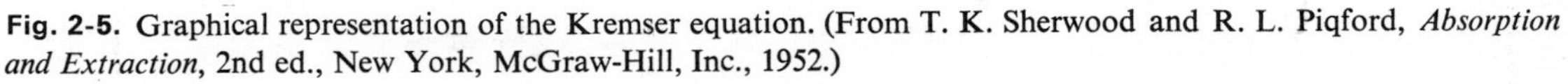
Fig. 2-5. Graphical representation of the Kremser equation. (From T. K. Sherwood and R. L. Piqford, *Absorption and Extraction*, 2nd ed., New York, McGraw-Hill, Inc., 1952.)

Sherwood and Pigford [2]. These graphs are an aid to visualizing the effects of N and E upon recovery, and they are also quite useful in providing an initial guess if Eq. (2-61) is to be solved by trial and error to obtain greater accuracy.

NOMENCLATURE

A = cross-sectional area of settling tank, ft^2
C_F = soda concentration in solution carried along with mud feed stream, moles of Na^+/liter
C_O = soda concentration in overflow stream, moles of Na^+/liter
C_U = soda concentration in solution in underflow stream, moles of Na^+/liter
d = diameter of settling tank, ft
E = extraction factor $\equiv W/SL$
e = 2.71828 . . .
f_j = fraction of wash water allocated to jth stage
L = liquid content of underflow or feed mud stream, volume of solution per mass of solid
N = number of stages in the cascade
R = fractional recovery of soda in washing system
S = solids feed rate, lb of solids/hr or tons of solids/day
W = wash-water flow rate, gal/min

GREEK LETTERS

λ = series defined by Eq. (2-57)
ρ_W = density of water, g/ml or lb/ft^3 or lb/gal
$\phi = 1/(1 - R)$

SUBSCRIPTS

j = refers to stage number j
N = refers to stage number N, the last stage

REFERENCES

1. A. KREMSER, *Natl. Petroleum News*, **22**, No. 21, 42 (May 21, 1930).
2. T. K. SHERWOOD and R. L. PIGFORD, *Absorption and Extraction*, 2nd ed., New York, McGraw-Hill, 1952, pp. 147, 407.

STUDY PROBLEMS

1. Consider the single-stage washing system shown in Fig. 2-2. Write a material-balance equation for each of the components—water, sodium ion, and "mud solids"—utilizing none of the assumptions described on pages 8 to 10. Carefully define additional symbols as needed. Show how these three equations are modified by each assumption in order to develop Eq. (2-1).

2. Repeat Example Problem 2-4 with the settling tanks swapped around so that L is equal to 4.5, 4.0, 3.5, and 5.0 ml of liquid per gram of solid in the underflow streams *leaving* the first, second, third, and fourth washing stages, respectively. The washing-system feed stream will have a liquid content of 3.0 ml/g, as in Example Problem 2-4.

3. The mud settler underflow stream in a proposed Bayer alumina plant will contain waste solids flowing at a rate of 1400 tons of solids per day and will have a liquid content of 3.8 ml/g of solid. If a four-stage countercurrent washing cascade is employed, at what rate must wash water be fed to the system to recover 90 per cent of the soda from the mud stream? Repeat the calculation for

Recovery, %	*Number of Stages*
90	10
98	4
98	10

4. A mud-washing system is to be designed to recover soda from a mud stream in a Bayer alumina plant. The feed stream to the mud-washing system will contain solid material flowing at a rate of 1600 tons of solids per day. The liquid content of the feed stream will be 4.2 ml of liquid per gram of solid, and the underflow streams from all washing stages will also have liquid contents of 4.2 ml/g. The liquid carried with the feed stream is 3.5 M in Na^+ ion content.

   Any soda lost to the river must be replaced in the process as sodium hydroxide, and thus the value of recovered soda is considered equivalent to sodium hydroxide at $0.03 per pound of NaOH. On the other hand, any *excess* wash water used in recovering the soda must be evaporated from the process (i.e., enough water must be evaporated so that the recovered soda is returned to the process at the feed concentration, C_F, which is 3.5 M in this case), and the cost of evaporating water is estimated to be $0.25 per ton of water. The cost of settling tanks is $30.00 per square foot of tank cross-sectional area. The capital investment required to build the mud washer is estimated to be 1.4 times the cost of the settling tanks; this will account for pumps, piping, and instrumentation. The size of the settling tanks is dictated by the settling velocity of the mud. With the present mud, the cross-sectional area of each settling tank must be sufficiently large so that the velocity calculated as the *total* volumetric flow rate of liquid fed to the tank divided by the tank cross-sectional area should not exceed 2 ft/hr. The rate of amortization of the capital investment will be 20 per cent of the invested capital per year, and this will account for depreciation, interest on invested capital, and maintanance costs. The operating labor costs will not be considered in this economic analysis because they are essentially fixed, independent of the design of the mud-washing system. An operating year of 363 days will be assumed.

(a) Based upon the simplified economic quide lines presented above, develop an equation relating the profit, P, from a countercurrent mud-washing operation to the number of stages, N, and E, the volumes of wash water used per volume of liquid carried with the mud stream. Express the profit in millions of dollars per year as the value of recovered soda minus the cost of evaporatint the excess wash water minus the amoritization of the capital investment in the washing system.

(b) Develop an analogous expression for the profit for a cross-flow mud-washing operation.

5. Write a computer program, using a simple search technique, to find the values of N and E that will maximize the profit for the countercurrent and cross-flow mud-washing systems described in Problem 4. Arrange the computer output to include the value of the recovered soda, the required capital investment, the cost of evaporation, and the profit for the optimum design. The output should also show these results at several values of E and N on either side of the optimum point to show how sharp or flat is the maximum in the profit function.

6. Referring to the alumina mud-washing system described in Problem 4, laboratory experiments have been conducted on the flocculation of the mud particles by the addition of strach. The results show that starch addition to the mud settler at a rate of 0.005 lb of starch per pound of mud solids will double the mud settling velocity. This will result in smaller settling tanks in the mud-washing system because the cross-sectional area of each tank can now be chosen such that the velocity calculated as the *total* volumetric flow rate of liquid fed to the tank divided by the tank cross-sectional area will not exceed 4 ft/hr (instead of 2 ft/hr as in Problem 4). Furthermore, the flocculated mud particles form a more compacted mud, with $L = 3.8$ ml of solution per gram of solid for all underflow streams and for the washer feed stream (instead of $L = 4.2$ ml/g as in Problem 4).

   Repeat Problems 4 and 5 for the washing system with starch addition at the rate of 0.005 lb per pound of mud solids. The cost of the starch at $0.05 per pound represents an additional operating expense to be subtracted from the profit function.

7. Referring to Problem 6, additional laboratory and pilot-plant experimentation have revealed that the flocculated mud is not repulpable. That is, the flocks are destroyed when the mud is redispersed in water, and additional starch is required to form them again. It now appears that the improvement in settling velocity and mud compaction described in Problem 6 can be achieved only by starch addition to the mud settler at a rate of 0.005 lb per pound of mud solids *plus* starch addition to *each stage* of the washing cascade at a rate of 0.003 lb per

pound of mud solids. Repeat Problem 6 with this new insight into the starch-addition requirements.

8. The profit functions required in Problems 4 through 7 were for the mud-washing operation alone and did not consider the primary mud settler, which settles the digester-effluent liquor into a recovered overflow stream and an underflow stream which is fed to the washing system. Maximizing these profit functions will yield optimum values of E and N for the mud-washing system for each case considered, but the advantage of adding starch cannot be appreciated by comparing directly the maximized profit values obtained in Problems 5, 6, and 7. The cost of the starch was charged to the mud-washing operation in the profit functions of Problems 6 and 7, but starch addition improves the operation of the primary mud settler as well as that of the washing system. First, the size of the primary mud settler is decreased by the starch addition because the mud settling velocity is increased. Second, the primary mud settler "recovers" more soda in its overflow stream when starch is added because the mud is more compact; therefore, less soda solution is fed to the mud-washing system. Clearly, the economic value of these advantages in the primary-mud-settler operation should be added to the profit function in Problems 6 and 7 to offset the cost of the starch.

   How much credit for improved primary mud settling should be added to the profit functions of Problems 6 and 7 to account for the decrease in mud-settler cost and for the improved soda recovery in the mud-settler overflow due to starch addition? The criterion for sizing the mud settler is the same as that used for the settling tanks in the washing system, based upon the mud-settling velocity. The 20 per cent per year amortization rate and the ratio of 1.4 for total investment to tank cost apply to the primary mud settler also. The digester effluent, which is the feed to the primary mud settler, contains 30 ml of soda solution per gram of mud solids.

Linear Cascades: Liquid–Liquid Extraction

3

Components in a liquid-phase solution may be separated by contacting that liquid with a second, partially miscible liquid in which the components to be separated have different solubilities. The degree of separation in a single contact is often quite small, but this can generally be amplified by means of a countercurrent cascade to produce any desired degree of separation. The two liquid phases will generally have different densities, and thus they can be separated from each other by settling under the influence of gravity or, more rapidly, by the use of a centrifugal field. The two liquid solvents must have limited solubility in each other so that two separate liquid phases will form. This generally requires that the liquids be of different chemical type or *polarity*. An example of a pair of liquids with very low mutual solubility would be water and a paraffin hydrocarbon such as octane. The highly polar water molecules and the nonpolar octane molecules dissolve in each other to less than 0.01 per cent at room temperature. When the two liquids are mixed and the mixture is allowed to settle under gravity, the more dense phase that settles to the bottom is essentially pure water with a very small amount of octane dissolved in it, while the less dense phase is essentially pure octane with a very small amount of water in it. If a third component, such as butyl alcohol, with a polarity intermediate between that of water and that of octane, is introduced into the system, it will distribute between the two liquid phases. If the system is agitated so that the phases are in intimate contact for a sufficiently long time, phase equilibrium can be closely approached.

LIQUID–LIQUID PHASE EQUILIBRIUM

For a given temperature the concentration of the distributing solute, such as butyl alcohol, in the organic phase will bear a unique relationship to its concentration in the aqueous phase, such as that shown qualitatively in Fig. 3-1. Generally speaking, unless the solute dissociates or polymerizes

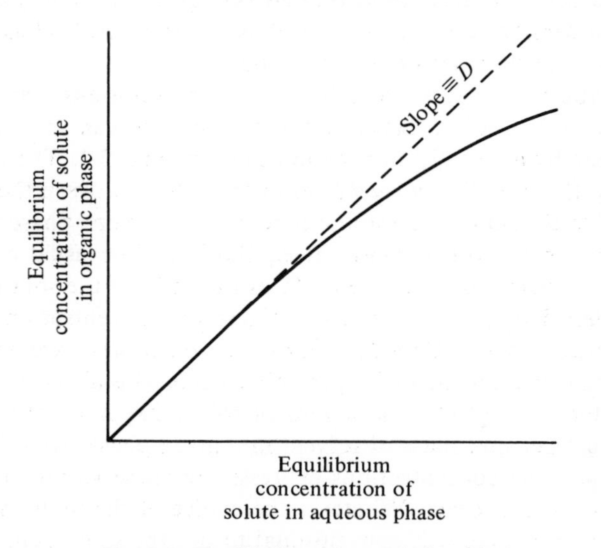

Fig. 3-1. Equilibrium-phase relationship.

in one liquid phase and not in the other, a plot of the equilibrium concentration of the solute in one phase versus the concerntration in the other phase will be linear for sufficiently dilute solutions. At high concentrations, departures from the linear relationship can become very appreciable. Furthermore, as more and more of the solute is added to the system, generally speaking, more and more of the water will dissolve in the octane phase and more and more of the octane will dissolve in the water phase. This is generally true, because if the solute has appreciable solubility in both water and in octane, large amounts of this solute in the water phase will result in an increase in the octane solubility in that phase, while large amounts of solute in the octane phase will result in appreciable solubility of water in that phase.

For many systems of this type, a point is reached where the concentration of the solute can be high enough to promote the mutual solubility of the solvents to the point where the entire system becomes a single-phase system, with all components dissolved in a single phase. Obviously, such high concentrations of the solute cannot be tolerated in a liquid–liquid extraction system, which is absolutely dependent upon the presence of two liquid phases that can be physically separated. Even before the solute concentration reaches the point where a single phase exists at equilibrium, as the concentration of the solute increases and becomes very substantial in both phases, the various important properties of the two phases tend to approach each other. For example, the density difference between the two phases will usually decrease steadily with increasing concentration of the solute and, therefore, with increasing mutual solubility of the two solvents. This is detrimental to the extraction process because, even though two phases may be present, if the difference in density between the two phases is too small, phase separation becomes costly and perhaps even impractical.

If attention is confined to relatively dilute solutions, the equilibrium distribution of the solute between the two phases can in most cases be approximated by a straight line, as indicated in Fig. 3-1. The slope of this line is called the *distribution coefficient* of the solute between the two phases and is equal to the ratio of the solute concentration in one phase to the solute concentration in the other phase, at equilibrium. The usual convention is to define the distribution coefficient, D, as the ratio of the solute concentration in the less dense, or *light*, *phase* to the solute concentration in the more dense, or *heavy*, *phase*. Concentrations are sometimes expressed as mole fractions and sometimes as moles per unit volume. It is also often convenient to express the concentration as moles of solute per mole of solvent or as mass of solute per unit mass of solvent in a given phase. At sufficiently low concentrations, the concentrations expressed in these various terms are all proportional to each other. Whatever the choice of the units in which concentration is expressed, it is conventional to use the same units for concentration in both phases. Here concentrations will be expressed in terms of mole ratios, moles of solute per mole of solvent in a given phase. In the light phase the mole ratio will be given the symbol Y; X will refer to the moles of solute per mole of solvent in the heavy phase. The distribution coefficient, D, given by

$$D \equiv (Y/X)_{\text{equilibrium}} \tag{3-1}$$

will be assumed to be constant in the mathematical description of the extraction process, but it will be kept in mind that this assumption will break down at sufficiently high concentrations of the distributing solute.

If a different distributing solute is added to the system consisting of, for example, octane and water, it will generally be expected that it will distribute between the two phases in a different ratio from that corresponding to, for example, butyl alcohol. Thus, if hexyl alcohol is added to a mixture of octane and water and the system is brought to equilibrium, one would expect the hexyl alcohol to favor distribution into the octane more than does butyl alcohol. The hexyl alcohol has an hydroxyl group, which is polar, but it has a longer hydrocarbon chain than does butyl alcohol, and thus it is a less polar molecule. Accordingly, its distribution coefficient, D, between water and octane would be expected to be substantially higher than the distribution coefficient of butyl alcohol between water and octane.

If both butyl alcohol and hexyl alcohol are added to a mixture of water and octane and the system is brought to equilibrium, each alcohol will be distributed between the two phases in approximately the same ratio as it would if the other alcohol were not present, provided that the alcohol concentrations are relatively low. At higher concentrations the presence of the alcohols in the two liquid phases will alter the properties of those phases and change not only the phase densities, but also the distribution coefficients of the two alcohols. Generally speaking, as the concentrations of the alcohols in the solvents increase, the densities of the two phases will approach each other and the distribution coefficients of the various solutes will approach unity. Finally, at sufficiently high solute concentrations, complete miscibility often results. In the mathematical models of the extraction process to be considered here, it will be assumed that the solute concentrations are sufficiently low that the distribution coefficients of the various solutes between the two phases are constant and independent of the presence of other solutes, but it must be remembered that this approximation breaks down at high concentrations of the solutes.

If two solutes, such as butyl and hexyl alcohols, are distributed at equilibrium between two phases, such as an aqueous phase and an octane phase, an important parameter which describes the degree of separation of the solutes by a single extraction step is the *relative distribution coefficient*, α, defined as

$$\alpha_{A,B} \equiv \left(\frac{Y_A/X_A}{Y_B/X_B}\right)_{\text{equilibrium}} \equiv \left(\frac{Y_A/Y_B}{X_A/X_B}\right)_{\text{equilibrium}} \tag{3-2}$$

The subscripts A and B refer to the two solute components and thus $\alpha_{A,B}$ is equal to the ratio of the moles of component A to the moles of component B in the light phase divided by the ratio of the moles of component A to the moles of component B in the heavy phase, at equilibrium. From the definition of the distribution coefficient, D, and the assumption that the distribution coefficient is constant, independent of composition, and independent of the presence of other solutes, it follows that the relative distribution coefficient is simply the ratio of the distribution coefficients for the two components

$$\alpha_{A,B} = \frac{D_A}{D_B} \tag{3-3}$$

The order of the subscripts on α indicates which distribution coefficient goes in the numerator, according to Eq. (3-3). The usual convention is to define α such that it is greater than unity, and thus component A would generally be the component with the greater distribution coefficient; if the subscripts are omitted from the α, this convention is assumed. The magnitude of the relative distribution coefficient, α, is a measure of the ease of separation of the two solutes. If α equals unity, no separation whatsoever is achieved because the two components distribute between the two phases in the same ratios. The greater the departure of α from unity, the greater the separation that can be achieved in a single contacting stage, and, generally speaking, the more attractive a liquid–liquid extraction process will appear.

The choice of the liquid solvents to be used is a very important consideration in the design of a liquid–liquid extraction system. A value of the relative distribution coefficient, α, that is substantially different from unity is highly desirable. It is also generally desirable that the individual distribution coefficients should not both be much less than unity or much greater than unity. For example, if two solutes have distribution coefficients equal to 2×10^{-4} and 1×10^{-4}, respectively, the relative distribution coefficient, α, equals 2, and this is a reasonable value for considering a liquid–liquid extraction separation process. But such low values of the individual distribution coefficients indicate that both solute components will scarcely be present in the light-solvent phase; thus enormous quantities of light solvent would have to be used to extract an appreciable quantity of either solute into the light solvent.

The factors that influence the values of the distribution coefficients and the thermodynamic theory of such equilibrium-phase relationships represent an important branch of physical chemistry, the physical chemistry of solutions. Knowledge in this field is vital for providing theoretical guidance to experimental programs for selecting the proper solvents to achieve a desired liquid–liquid extraction separation. This subject will not be considered here; the equilibrium-phase behavior will be regarded as having been measured experimentally, and the focus here will be upon the behavior of the extraction cascade. But it should be kept in mind that the physical chemistry of solutions is an important field of knowledge required for an understanding of the equilibrium relationships in systems of this type.

SIMPLE EXTRACTION OF ONE SOLUTE

In many situations it is desired to extract a single solute component from one liquid-phase solution into another, immiscible, solvent. In this case, the solute to be extracted would distribute to an appreciable extent into the other solvent, while the other components in solution with it would distribute to a negligible extent. An example of this type of extraction process is the extraction of penicillin [19] from the aqueous fermentation broth into amyl acetate. Subsequently, the penicillin is extracted from the amyl acetate into an aqueous buffer solution. Another commercial example is the recovery of isobutylene [7] from solutions containing normal butenes and butanes by extraction into aqueous sulfuric acid solution. Here advantage is taken of a chemical reaction between the sulfuric acid and the isobutylene to form isobutyl sulfate, which is soluble in the aqueous phase. It is not uncommon to find extraction processes that depend upon a chemical reaction such as this, and it should be realized that the equilibrium-phase relationships will generally be more complex in such cases. Another example is the removal of aromatic hydrocarbons from kerosines [18] to improve the burning qualities of the kerosines. An early process due to Edeleanu accomplished this by extraction of the aromatic compounds into liquid sulfur dioxide. This solvent extracts the aromatic compounds quite selectively while leaving paraffinic and naphthenic compounds in the kerosine phase.

3-1. Single-stage extraction

Consider a heavy solvent containing a solute to be removed from that solvent phase by extraction into a light-solvent phase. This might be accomplished in a single-stage mixer–settler extraction unit, such as that shown in Fig. 3-2. The system could be operated batchwise, but continuous operation

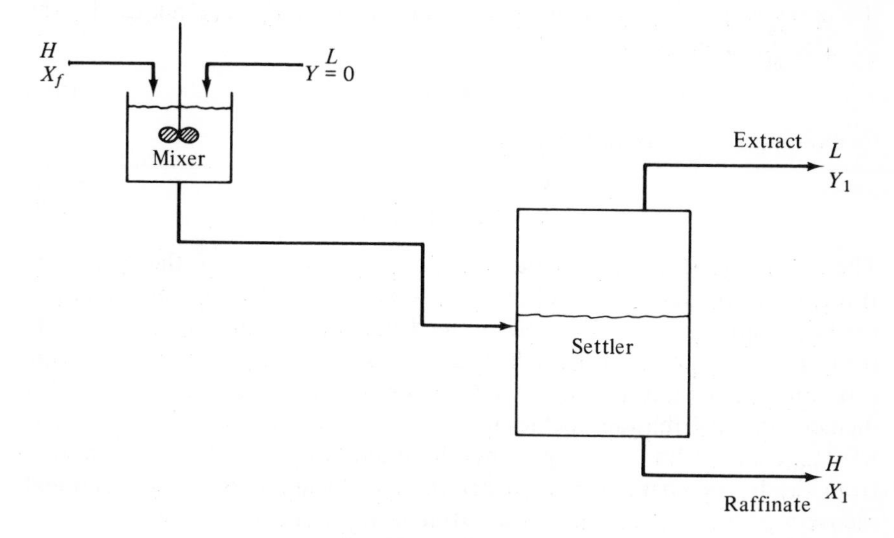

Fig. 3-2. Single-stage extraction unit.

will be considered here. The heavy-solvent phase flows continuously into the mixer at a rate of H moles of heavy solvent per hour; the light solvent flows steadily into the mixer at a rate of L moles per hour. The solute content of the heavy-phase feed is X_f moles of solute per mole of heavy solvent, and the light-solvent feed will be assumed to be free of the solute. In the mixer the two immiscible phases are agitated and broken into small droplets, facilitating the diffusion of the solute species from the heavy phase to the light phase. The discharge from the mixer will generally be a dispersion of droplets of one phase in the other phase. This stream will flow steadily to the settler, where the cross-sectional area is sufficiently large that the liquid velocity will be quite small, and the two phases can separate under the influence of gravity. The overflow from the settler is the light phase containing the extracted solute, and this stream is usually referred to as the *extract*. The underflow from the settler is the heavy solvent now stripped of some of its solute content. This stream is commonly called the *raffinate* stream.

If the heavy solvent and the light solvent can be considered to be essentially insoluble in each other, the molal flow rate of heavy solvent in the raffinate stream will be equal to that in the heavy-solvent feed stream, H moles of heavy solvent per hour. Likewise, the flow rate of light solvent in the extract stream will be equal to the feed rate of light solvent, L moles of light solvent per hour. A material balance on the solute can be written as

$$HX_f = HX_1 + LY_1 \tag{3-4}$$

Assuming sufficiently intimate mixing of the phases and sufficiently long residence time in the mixer so that the phases closely approach equilibrium, the compositions of the two phases leaving the settler are related by the equilibrium expression

$$Y_1 = DX_1 \tag{3-5}$$

Combining Eqs. (3-4) and (3-5) yields

$$\frac{X_1}{X_f} = \frac{1}{1 + (LD/H)} \tag{3-6}$$

The group LD/H is called the *extraction factor* and is given the symbol E. It is equal to the rate of flow of the light solvent divided by the rate of flow of the heavy solvent and multiplied by the distribution coefficient of the solute from the heavy phase to the light phase. It is a simple generalization of the extraction factor introduced in Chap. 2 for the mud-washing system, but it includes the distribution coefficient, D, as well as the flow-rate ratio. The left-hand side of Eq. (3-6) represents the fraction of the solute not removed from the heavy phase in the extraction step. Defining R as the fractional recovery of the solute in the extract stream, Eq. (3-6) can be written

$$1 - R = \frac{1}{1 + E} \tag{3-7}$$

which can be seen to be identical with Eq. (2-4) for a single-stage mud-washing system. As in the mud-washing system, the recovery can be made as high as desirable by simply increasing L, the moles of light solvent fed to the extraction system. This is done, however, at the expense of recovering the solute at a very low concentration in the light-solvent phase. The amount of light solvent required can be decreased, and thus the concentration of the recovered solute in the extract can be increased by using a multistage cascade of the cross-flow type or, even more efficiently, of the countercurrent type. The analysis of these systems is a straightforward extension of that presented in Chap. 2, the only change being that the extraction factor, E, includes the distribution coefficient in its definition.

3-2. Countercurrent extraction

For example, consider the countercurrent extraction system shown in Fig. 3-3. This is a schematic diagram with each stage represented by a simple

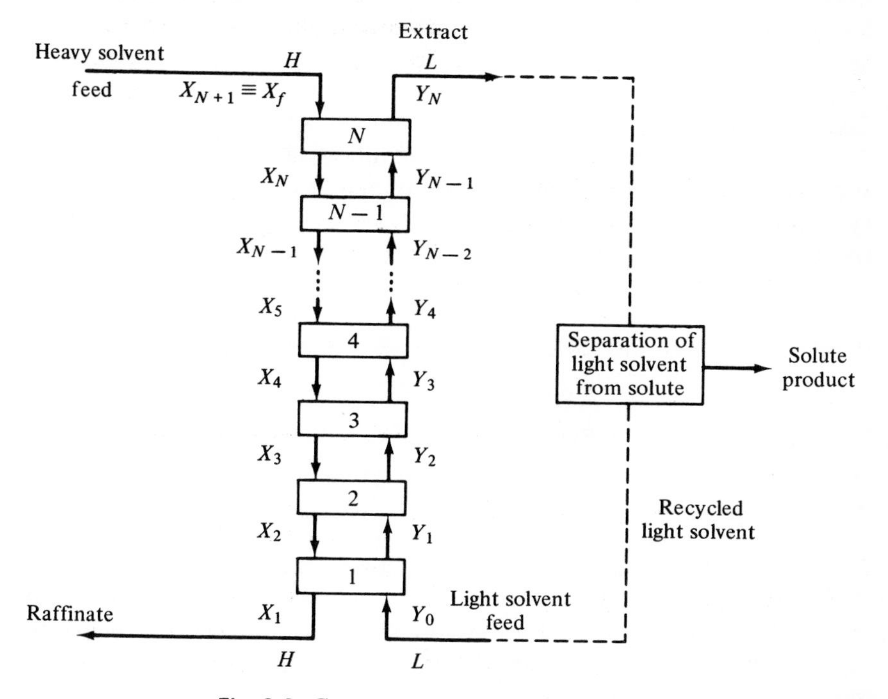

Fig. 3-3. Countercurrent extraction system.

block meant to represent the contacting stage, including the mixer and the settler or whatever the contacting device may be. The block simply represents the equipment in which a given light stream and a given heavy stream enter and are mixed such that they are brought to equilibrium; thus the light and heavy streams leaving the stage are assumed to be in equilibrium if an "ideal stage" is assumed. Note that the streams entering any stage are the light-solvent stream from the stage beneath plus the heavy-solvent stream from the stage above. This is completely analogous to the countercurrent washing system shown in Fig. 2-4.

Assume that the heavy-solvent stream contains the solute component at a concentration of X_f moles of solute per mole of heavy solvent and that it is desired to extract this solute into the light-solvent stream. For the sake of generalizing somewhat, this time the light solvent will not be assumed to enter the system solute free; rather the light solvent will be assumed to contain a small amount of solute as it enters, Y_0 moles of solute material per mole of light solvent. This is not an unrealistic assumption because in most extraction processes such as this, the extract phase of light solvent leaving the Nth stage and containing the extracted solute component at the concentration Y_N would generally be sent to another separation system, where the desired solute and the light solvent would be separated, perhaps by vaporizing the light solvent away from the solute. After separation, the light solvent would be recycled to the first stage. Often the separation of the solute from the light solvent will be incomplete, and thus the light-solvent feed to the system will contain a small amount of the solute.

The performance of the countercurrent extraction cascade shown in Fig. 3-3 can be described by an equation similar to Eq. (2-61) for the countercurrent mud-washing system discussed in Chap. 2. A material balance for the solute component around the first stage yields

$$HX_2 + LY_0 = HX_1 + LY_1 \tag{3-8}$$

Similarly, material balances around the second and third stages yield

$$HX_3 + LY_1 = HX_2 + LY_2 \tag{3-9}$$

$$HX_4 + LY_2 = HX_3 + LY_3 \tag{3-10}$$

All these equations take the form, for the jth stage,

$$HX_{j+1} + LY_{j-1} = HX_j + LY_j \tag{3-11}$$

Equation (3-11) applies to all stages, including the Nth stage, provided that X_{N+1} is identified with X_f, the solute content of the heavy-solvent feed stream. Assuming that the light- and heavy-solvent streams leaving any stage are in equilibrium, the Y terms can be eliminated by using the equilibrium expression

$$Y_j = DX_j \qquad (j = 1, 2, \ldots, N) \tag{3-12}$$

and these equations can be written as

$$X_2 - X_1 = E(X_1 - X_0) \tag{3-13}$$

$$X_3 - X_2 = E(X_2 - X_1) \tag{3-14}$$

$$X_4 - X_3 = E(X_3 - X_2) \tag{3-15}$$

$$\vdots$$

$$X_{j+1} - X_j = E(X_j - X_{j-1}) \tag{3-16}$$

As before, the extraction factor is defined as

$$E \equiv \frac{LD}{H} \tag{3-17}$$

The symbol X_0 appearing in Eq. (3-13) deserves special mention. In transforming Eq. (3-8) to the form of Eq. (3-13), it was convenient to define X_0 as

$$X_0 \equiv \frac{Y_0}{D} \tag{3-18}$$

to preserve the symmetry of Eqs. (3-13) through (3-16). Thus the term X_0 simply represents the solute content of the light solvent fed to the first stage divided by the distribution coefficient. Therefore, X_0 does not represent an actual composition of a heavy-solvent stream; it represents the solute content of a fictitious heavy-solvent stream which would be in equilibrium with the light-solvent feed to the first stage. Equations (3-13) through (3-16) are identical with the system of Eqs. (2-46) through (2-49) for the countercurrent mud-washing system except for the more general definition of the extraction factor, E, and the presence of the term X_0, representing the solute content of the light-solvent feed in the present case.

Proceeding in the same way in which Eq. (2-53) was derived from Eqs. (2-46) through (2-49), Eqs. (3-13) through (3-16) can be manipulated to yield

$$X_{j+1} - X_1 = (E + E^2 + E^3 + \cdots + E^j)(X_1 - X_0) \tag{3-19}$$

in general, or, for $j = N$,

$$X_f - X_1 = (E + E^2 + E^3 + \cdots + E^N)(X_1 - X_0) \tag{3-20}$$

Equation (3-20) can be rewritten as

$$X_f - X_0 = (1 + E + E^2 + E^3 + \cdots + E^N)(X_1 - X_0) \tag{3-21}$$

and, using Eq. (2-56), as the more general Kremser [6] equation,

$$\frac{X_1 - X_0}{X_f - X_0} = \frac{1 - E}{1 - E^{N+1}} \tag{3-22}$$

For the special case in which X_0 is equal to zero, the left-hand side of Eq. (3-22) equals the fraction of the solute material in the heavy-solvent feed not extracted into the light-solvent stream in the countercurrent cascade;

in this case, Eq. (3-22) is identical with Eq. (2-61) for a mud-washing cascade with soda-free wash water. [If the wash water in the mud-washing cascade of Chap. 2 contained an appreciable soda content, Eq. (2-61) would have to be modified to a form analogous to that of Eq. (3-22).] In the general case in which X_0 is not equal to zero, the left-hand side of Eq. (3-22) does not represent the fraction of the solute material not extracted, but, knowing the values of X_0 and X_f, Eq. (3-22) can be used to calculate the fractional recovery.

The preceding derivation was approached from the viewpoint of extracting a solute material from the heavy-solvent stream into the light-solvent "wash liquid." But, since generality was maintained by allowing the light-solvent feed stream to contain the solute at a concentration Y_0, Eq. (3-22) may be used for a system in which the solute content of the light-solvent feed is to be extracted into a heavy-solvent "wash liquid." The overall material balance around the entire countercurrent extraction cascade is written

$$H(X_f - X_1) = L(Y_N - Y_0) \tag{3-23}$$

Eliminating X_1 between Eqs. (3-22) and (3-23) and using Eq. (3-17) yields, after some manipulation,

$$\frac{Y_N - Y_f}{Y_0 - Y_f} = \frac{1 - 1/E}{1 - 1/E^{N+1}} \tag{3-24}$$

The term Y_f is defined as

$$Y_f \equiv DX_f \tag{3-25}$$

and thus it corresponds to the solute content of a fictitious light-solvent stream that would be in equilibrium with the heavy-solvent feed stream. If Y_f equals zero, the left-hand side of Eq. (3-24) represents the fraction of the solute content of the light-solvent feed not extracted into the heavy solvent in the countercurrent extraction cascade.

Indeed, Eqs. (3-22) and (3-24) are completely symmetrical, the former adopting the point of view of extracting a solute from the heavy-solvent stream into the light-solvent stream and the latter adopting the point of view of extracting a solute from the light-solvent stream into the heavy-solvent stream. The extraction factor, E, is a measure of how readily the light-solvent stream can remove solute from the heavy-solvent stream. Likewise, the reciprocal of the extraction factor is a measure of how readily the heavy-solvent stream can extract the solute from the light-solvent stream. Looked at another way, the greater the value of E the greater is the tendency of the solute component to be washed to the top of the cascade by the upward-flowing light-solvent stream; the smaller the value of E the greater is the tendency of the solute component to be washed to the bottom of the cascade by the downward-flowing heavy-solvent stream.

For any fixed value of E greater than unity, it is apparent from Eq. (3-22) that as N tends to infinity, X_1 approaches X_0. Thus the composition of the heavy stream leaving the cascade approaches that value that would correspond to equilibrium with the light-solvent stream entering the cascade. For any fixed value of E less than unity, it is apparent from Eq. (3-24) that as N approaches infinity, Y_N approaches Y_f. In this case, the solute content of the light-solvent stream leaving the cascade approaches a value that corresponds to equilibrium with the heavy-solvent stream entering the cascade. Thus, as the number of stages in the cascade tends toward infinity, equilibrium is approached at the bottom of the cascade or at the top of the cascade, depending upon whether E is greater than or less than unity. For the special case of E equal to unity, Eqs. (3-22) and (3-24) become

$$\frac{X_1 - X_0}{X_f - X_0} = \frac{Y_N - Y_f}{Y_0 - Y_f} = \frac{1}{N+1} \qquad \text{(for } E = 1\text{)} \qquad (3\text{-}26)$$

In this case, as the number of stages tends toward infinity, the compositions of the two streams approach equilibrium with each other at the top of the cascade and at the bottom of the cascade and, indeed, throughout the entire cascade.

Figure 2-5 can obviously be regarded as a solution to Eq. (3-22) in which the left-hand side of this equation is represented by $(1 - R)$ in the figure, although in this sense R would not equal recovery unless X_0 were equal to zero. In a similar manner, Fig. 2-5 could be used to obtain solutions to Eq. (3-24) by identifying E with its reciprocal.

In Eqs. (3-1) through (3-26), the solvent flow rates, L and H, have been expressed as moles of solvent per unit time, and the concentrations, Y and X, have been expressed as moles of solute per mole of solvent in a given phase. It is equally valid to define L and H as the mass of solvent flowing per unit time while defining Y and X as the mass of solute per unit mass of solvent in a given phase. In this case a product LY or HX would represent a mass flow rate of the solute in a given phase. The basic material balances, such as Eqs. (3-4) and (3-8) through (3-11), would then be equally valid provided that a consistent set of units were adopted.

EXAMPLE PROBLEM 3-1 Linoleic acid ($C_{17}H_{31}COOH$) has a distribution coefficient of 2.17 between the light-solvent heptane and the heavy-solvent methyl Cellosolve (ethylene glycol methyl ether) containing 10 per cent water by volume. This distribution coefficient is based upon Y and X defined as mass of solute per unit mass of solvent. A stream of the heavy solvent containing 0.1 lb of linoleic acid per pound of heavy solvent is to be scrubbed with a pure light-solvent stream, containing no linoleic acid, for the purpose of extracting 98 per cent of the linoleic acid from the heavy-solvent stream into the light-solvent stream. What is the required ratio of light-solvent to heavy-solvent flow rates if a ten-stage countercurrent extraction system is used?

Solution: Since the light solvent contains no linoleic acid, $X_0 = 0$. With $N = 10$ and a 98 per cent recovery, Eq. (3-22) becomes

$$0.02 = \frac{E - 1}{E^{11} - 1}$$

By trial and error the solution is found to be

$$E = 1.279$$

Since $D = 2.17$, the light-solvent requirement is

$$\frac{L}{H} = \frac{1.279}{2.17} = 0.589 \text{ lb of light solvent/lb of heavy solvent}$$

EXAMPLE PROBLEM 3-2 Repeat Example Problem 3-1 assuming that the light-solvent feed contains 0.001 lb of linoleic acid per pound of light solvent.

Solution: Since $Y_0 = 0.001$,

$$X_0 = \frac{0.001}{2.17} = 0.00046$$

employing Eq. (3-18). Since $X_f = 0.1$, for 98 per cent recovery $X_1 = 0.002$. Thus Eq. (3-22) becomes

$$\frac{E - 1}{E^{11} - 1} = \frac{0.002 - 0.00046}{0.1 - 0.00046} = 0.0155$$

By trial and error the solution is found to be

$$E = 1.324$$

and thus the light-solvent requirement is

$$\frac{L}{H} = \frac{1.324}{2.17} = 0.610 \text{ lb of light solvent/lb of heavy solvent}$$

EXAMPLE PROBLEM 3-3 Repeat Example Problem 3-1 assuming that the light solvent contains 0.006 lb of linoleic acid per pound of light solvent.

Solution: Since $Y_0 = 0.006$,

$$X_0 = \frac{0.006}{2.17} = 0.00277$$

As in Example Problem 3-2, $X_f = 0.1$ and $X_1 = 0.002$. Thus Eq. (3-22) becomes

$$\frac{E - 1}{E^{11} - 1} = \frac{0.002 - 0.00277}{0.1 - 0.00277} = -0.00791$$

There is no real positive value of E that will satisfy this equation, and it is apparent that this extraction cannot be accomplished with any amount of this light solvent. As E approaches infinity with $N = 10$, it is seen from Eq. (3-22) that X_1 approaches X_0, which corresponds to equilibrium with the light-solvent feed stream. Therefore, extraction with this light-solvent stream can reduce the linoleic acid content of the heavy solvent only to a value of $X_1 = 0.00277$, which corresponds to 97.23 per cent removal of the linleic acid from the heavy-solvent stream.

EXAMPLE PROBLEM 3-4 A light-solvent stream (heptane) contains 0.1 lb of linoleic acid per pound of solvent. It is desired to remove 98 per cent of this linoleic acid by extraction into a heavy-solvent stream (methyl Cellosolve containing 10 per cent water by volume). The heavy solvent contains no linoleic acid. What is the required ratio of heavy solvent to light solvent if 10 stages are used in the countercurrent cascade?

Solution: From the symmetry of Eqs. (3-22) and (3-24) and from the solution to Example Problem 3-1, it is apparent that

$$\frac{1}{E} = 1.279$$

Thus, the heavy-solvent requirement is given by

$$\frac{H}{L} = (1.279)(2.17) = 2.78 \text{ lb of heavy solvent/lb of light solvent}$$

3-3. Graphical representation of the countercurrent cascade

Consider a material balance around the first j stages of a countercurrent cascade as depicted by the cascade in Fig. 3-4 and the material-balance envelope indicated by the dashed line. The material balance for the solute component is

$$H(X_{j+1} - X_1) = L(Y_j - Y_0) \tag{3-27}$$

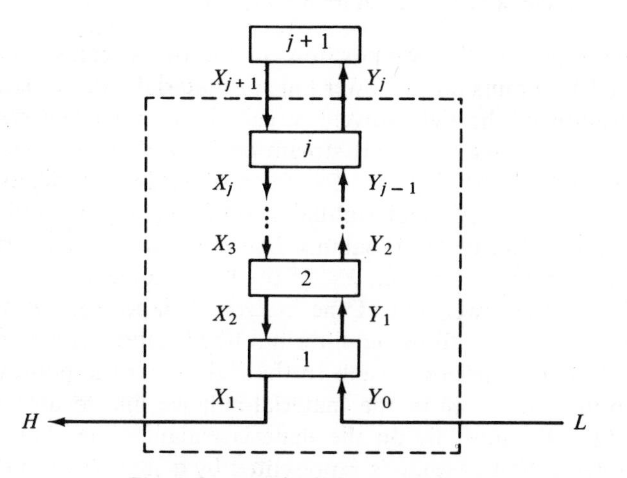

Fig. 3-4. Material-balance envelope.

If this material-balance equation is plotted as Y_j versus X_{j+1}, it is represented by a straight line with a slope H/L which passes through the point (X_1, Y_0). An example of such a material-balance line is shown as the lower straight line in Fig. 3-5. Assuming that the stages in the cascade are ideal stages, a plot of Y_j versus X_j, according to Eq. (3-12), is a straight line through the origin with a slope equal to D. This is shown as the upper straight line in the example in Fig. 3-5. Points on the upper line, or equi-

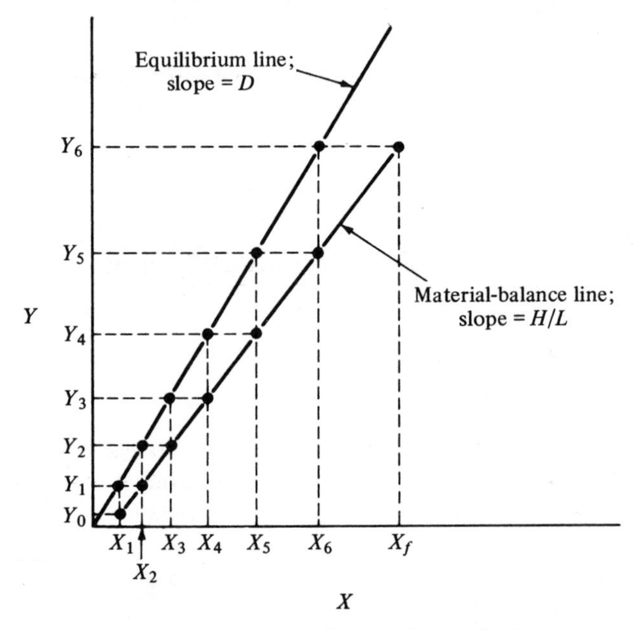

Fig. 3-5. Extraction from heavy solvent, six stages.

librium line, represent the compositions of the two streams leaving a given stage, X_j and Y_j. Points on the lower line, or material-balance line, represent the composition of the light-solvent stream leaving a given stage and the composition of the heavy-solvent stream entering that same stage.

The example shown in Fig. 3-5 is for a six-stage cascade, and it is seen that six "steps" from the material-balance line up to the equilibrium line and then over to the material-balance line connect the exit concentration, X_1, with the feed concentration, X_f, of the heavy-solvent stream. The point (X_1, Y_0) lies at the lower end of the material-balance line. A vertical step upward from this point to the equilibrium line locates Y_1, which is in equilibrium with X_1. A horizontal step to the right from the point (X_1, Y_1) on the equilibrium line over to the material-balance line locates X_2, because the point (X_2, Y_1) must lie on the material-balance line. Thus each stage in the countercurrent cascade is represented by a step between the material-balance line and the equilibrium line. In the example in Fig. 3-5, the material is being extracted from the heavy-solvent stream into the light-solvent stream, and thus X_1 is smaller than X_f. It should be noted that E is the ratio of the slope of the equilibrium line to the slope of the material-balance line, and in the example shown in Fig. 3-5 this ratio is greater than unity.

Figure 3-6 shows an example of a six-stage countercurrent cascade

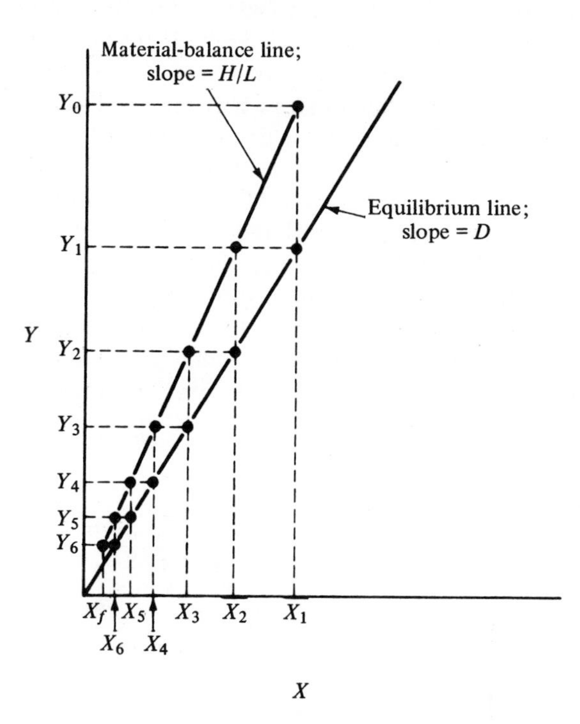

Fig. 3-6. Extraction from light solvent, six stages.

in which the solute is being extracted from the light phase into the heavy phase; thus the exit concentration, X_1, of the heavy-solvent phase is greater than the feed concentration, X_f. In this case, the material-balance line lies above the equilibrium line. Starting from the point (X_1, Y_0), which represents the terminal concentrations at the bottom of the cascade, a vertical step downward to the equilibrium line locates Y_1. From this point a horizontal step to the left over to the material-balance line locates X_2. Successive steps down to the equilibrium line and over to the left to the material-balance line locate the successive concentrations up the cascade until finally the terminal concentrations X_f and Y_6 are located. Again, each stage in the cascade can be represented by a step from the material-balance line to the equilibrium line and back to the material-balance line.

When plotted in the coordinates of Figs. 3-5 and 3-6, the graphical representation will always have the material-balance line below the equilibrium line when the solute is being extracted from the heavy solvent into the light solvent and will have the material-balance line above the equilibrium line when the solute is being extracted from the light solvent into the heavy solvent. In the example shown in Fig. 3-6, E, the ratio of the slope of the equilibrium line to that of the material-balance line, is less than unity.

As the number of stages in the cascade tends toward infinity, it is clear that the two lines must approach each other very closely at some *pinch point* in order that an infinite number of steps between the lines should result in a finite change in composition. For extraction from the light solvent into the heavy solvent with E less than unity, such as in Fig. 3-6, the pinch point takes place at the top of the cascade (but at the bottom of the diagram in Fig. 3-6). Figure 3-7 shows the graphical representation of a countercurrent cascade with the same equilibrium line, the same *slope* of the material-balance line, and the same feed compositions Y_0 and X_f as those shown in Fig. 3-6; but the cascade depicted in Fig. 3-7 has a greater number of stages, and thus the material-balance line pinches in on the equilibrium line at the lower left-hand corner of the diagram. This results in a somewhat

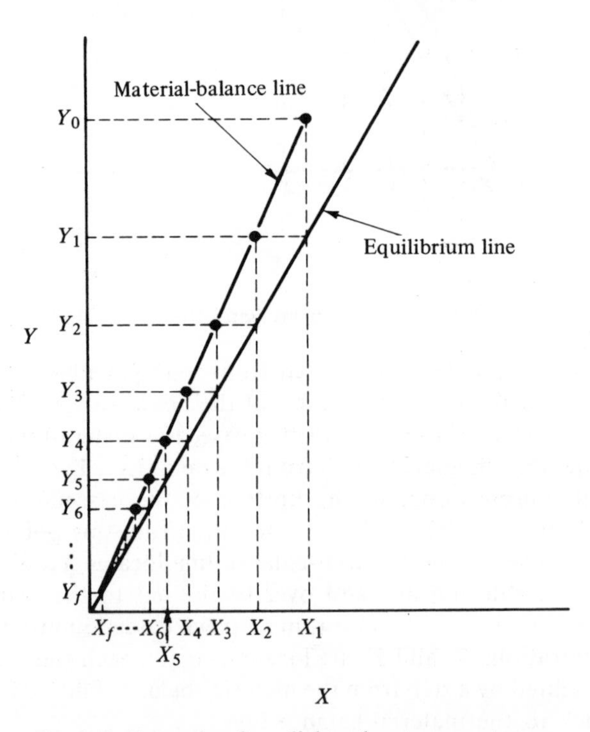

Fig. 3-7. Extraction from light solvent, many stages.

smaller value of Y from the final stage and a somewhat larger value of X_1 than the values shown in Fig. 3-6. As the number of stages tends toward infinity, it is evident in Fig. 3-7 that Y_N approaches Y_f, the value of Y in equilibrium with X_f, a conclusion that was reached earlier by examining Eq. (3-24).

Graphical constructions such as those in Figs. 3-5 through 3-7 are useful supplements to Eqs. (3-22) and (3-24) for the design of linear countercurrent extraction cascades because they provide a means of visualizing the effects of various changes in the system parameters. Furthermore, in a more general case when the equilibrium line is nonlinear, Eqs. (3-22) and (3-24) are not applicable, but the graphical construction is still valid. These diagrams are especially useful in binary distillation problems, where they were first introduced by McCabe and Thiele [8] and are commonly called McCabe–Thiele diagrams.

EXAMPLE PROBLEM 3-5 Abietic acid ($C_{19}H_{29}COOH$) distributes between the light-solvent heptane and the heavy-solvent methyl Cellosolve containing 10 per cent water by volume with a distribution coefficient of 1.57, Y and X being defined as mass ratios. A stream of this heavy solvent, containing 0.1 lb of abietic acid per pound of heavy solvent, is to be extracted with the light solvent in a 10-stage countercurrent extraction system with a light-solvent to heavy-solvent flow-rate ratio of 0.589 lb of light solvent per pound of heavy solvent. If the light-solvent feed to the cascade is free of abietic acid, what fraction of the abietic acid in the heavy-solvent feed will be extracted into the light solvent?

Solution: The extraction factor for abietic acid is

$$E = \frac{LD}{H} = (0.589)(1.57) = 0.925$$

Substitution into Eq. (3-22) yields

$$\frac{X_1}{X_f} = 1 - R = \frac{1 - 0.925}{1 - (0.925)^{11}} = 0.13$$

Thus 87 per cent of the abietic acid is extracted from the heavy phase into the light phase.

EXAMPLE PROBLEM 3-6

(a) A heavy-solvent stream of methyl Cellosolve containing 10 per cent water by volume contains 0.05 lb of linoleic acid and 0.05 lb of abietic acid per pound of heavy solvent. If this stream is extracted with a pure heptane stream in a 10-stage countercurrent extraction cascade with a solvent ratio of 0.589 lb of light solvent per pound of heavy solvent, what percentage of each of the acids will be extracted into the light-solvent phase?

(b) Referring to the problem of part a, what value of the light-solvent to heavy-solvent flow-rate ratio and what value of the number of stages do you recommend for effecting a good separation between the linoleic and abietic acids—that is, for achieving a separation in which most of the linoleic acid leaves the cascade with the light-solvent phase and most of the abietic acid leaves with the heavy-solvent phase?

Solution:

(a) Assuming that at these dilute concentrations the phase equilibrium can be described by constant distribution coefficients of 2.17 and 1.57 for linoleic and abietic acids, respectively, the extraction of each acid from the heavy solvent into the light solvent will be independent of the presence of the other acid. From the solutions to Example Problems 3-1 and 3-5, it follows that 98 per cent of the linoleic acid and 87 per cent of the abietic acid in the heavy-solvent feed will be extracted into the light-solvent extract.

(b) No values of L/H and N can be chosen that will effect a good separation between the linoleic and abietic acids. Consider first a cascade with N approaching infinity, which should yield the best results. Equation (3-22) takes the form

$$R = E \quad \text{if} \quad E < 1$$

$$R = 1 \quad \text{if} \quad E > 1$$

where R is the fraction of a solute component extracted from the heavy phase into the light phase. This limiting form of Eq. (3-22) as N approaches infinity was presented in Eq. (2-63) and the discussion following. Denoting linoleic and abietic acids as components A and B, respectively,

$$E_A = 2.17\left(\frac{L}{H}\right)$$

$$E_B = 1.57\left(\frac{L}{H}\right)$$

because $D_A = 2.17$ and $D_B = 1.57$. Combining these equations yields

$$R_A = 2.17\left(\frac{L}{H}\right) \quad \text{for } \frac{L}{H} < 0.461$$

$$R_A = 1 \quad \text{for } \frac{L}{H} > 0.461$$

and

$$R_B = 1.57\left(\frac{L}{H}\right) \quad \text{for } \frac{L}{H} < 0.637$$

$$R_B = 1 \quad \text{for } \frac{L}{H} > 0.637$$

These equations are plotted as the straight lines in Figs. 3-8 and 3-9 as R versus L/H for each component. It is seen that no value of L/H gives simultaneously a high value of R_A and a low value of R_B. For example, at $L/H = 0.267$, $R_A = 0.58$ and $R_B = 0.42$. Thus 58 per cent of the linoleic acid but only 42 per cent of the abietic acid would be extracted into the light solvent. This would effect some degree of separation between the two acids, but the separation could hardly be called a good one.

Increasing L/H to 0.461 will increase R_A to unity and R_B to 0.724. Therefore, the heavy phase leaving the cascade would contain no linoleic acid, but it would

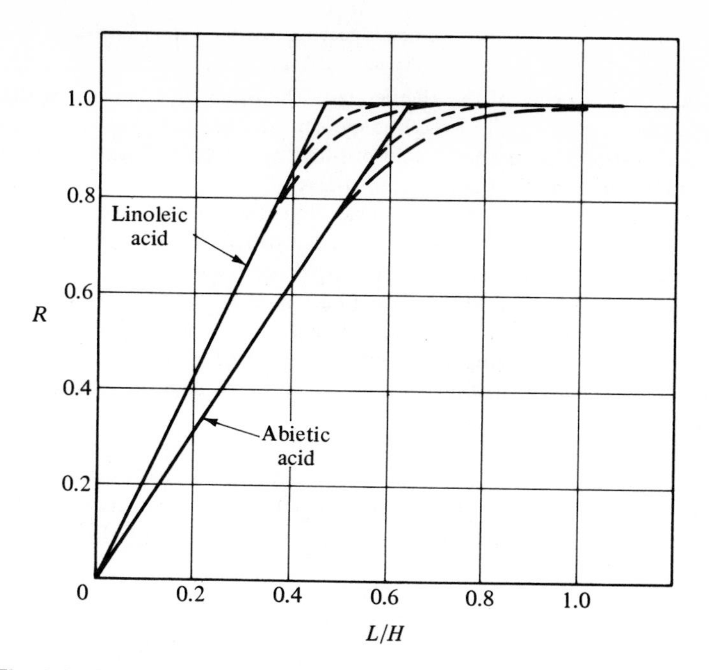

Fig. 3-8. Cascade performance for $N = 10$, — — —; $N = 17$, - - - -; and $N \longrightarrow \infty$, ————.

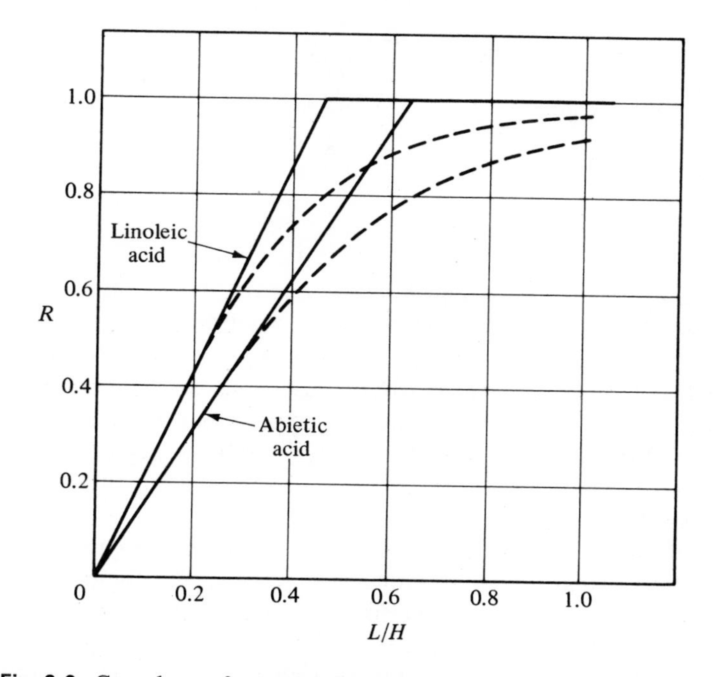

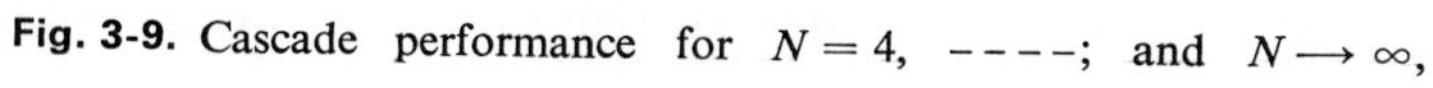

Fig. 3-9. Cascade performance for $N = 4$, - - - -; and $N \longrightarrow \infty$, ————.

contain only 27.6 per cent of the abietic acid. The other 72.4 percent of the abietic acid would be mixed with all of the linoleic acid in the light phase leaving the cascade. This will yield a pure abietic acid product but at only 27.6 per cent recovery, and the linoleic acid product is not at all pure.

Inspection of Figs. 3-8 and 3-9 shows that there is no value of L/H that will give, for example, $R_A = 0.99$ and $R_B = 0.01$, which would correspond to a 99 per cent recovery of each acid at a purity of 99 per cent.

For values of N less than infinity, Eq. (3-22) or Fig. 2-5 can be used to compute R versus L/H for each solute component, and results for $N = 4$, 10, and 17 are shown in Figs. 3-8 and 3-9 for comparison with the result for $N \longrightarrow \infty$. It is seen that the general conclusion is the same for other values of N. A good separation between the two acids cannot be achieved.

Reflection on these results shows that a good separation between two solute components can be achieved in a simple countercurrent extraction cascade *only* if their distribution coefficients differ by a large factor. Choosing

$$\frac{L}{H} = \frac{1}{D_A + D_B}$$

gives

$$E_A = \frac{D_A}{D_A + D_B} = \frac{\alpha}{\alpha + 1}$$

$$E_B = \frac{D_B}{D_A + D_B} = \frac{1}{\alpha + 1}$$

For $N \longrightarrow \infty$, this gives

$$R_A = \frac{\alpha}{\alpha + 1}$$

$$R_B = \frac{1}{\alpha + 1}$$

Thus, if $\alpha \equiv D_A/D_B$ is equal to 99, $R_A = 0.99$ and $R_B = 0.01$, and a sharp separation is achieved. For $\alpha = 49$, $R_A = 0.98$ and $R_B = 0.02$, which may also be called a good separation for many purposes. But with $\alpha = 4$, $R_A = 0.8$ and $R_B = 0.2$, which would leave each acid contaminated with 20 per cent of the other acid and would generally not be considered a good separation. Note that $R_A = 0.999$ and $R_B = 0.001$, a degree of separation often required in industrial practice, could be achieved only if α were at least as great as 999.

Therefore, it is seen that a simple countercurrent extraction cascade cannot effect a sharp separation between two solute components unless their relative distribution coefficient is quite large, of the order of 100. When α is close to unity, a sharp separation can be achieved by using a *fractional* extraction cascade, as described in a subsequent section of this chapter.

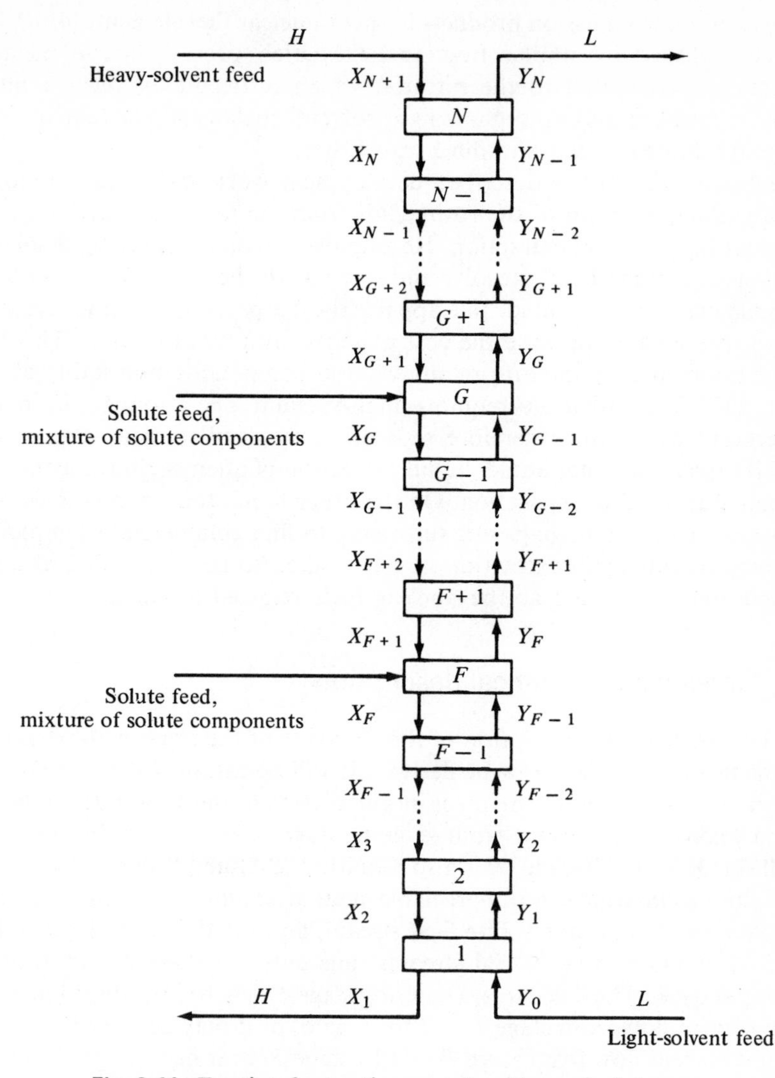

Fig. 3-11. Fractional-extraction cascade with two feed streams.

large amount of component A along with component B back to the feed stage. With a large number of stages in the lower section, removal of component A from the bottom product can be nearly complete. Simultaneously, the removal of component B from the top product can be nearly complete if the upper section of the cascade contains a large number of stages.

Commercial examples of the use of fractional-extraction cascades are found in the nuclear power industry. Uranium-metal purification [1, 3, 5], the separation of zirconium from hafnium [3], and the separation of uranium and plutonium from fission products in spent nuclear fuel elements [4, 10, 21] are accomplished by similar fractional-extraction processes. The metallic elements are converted to the nitrates, which distribute between a nitric acid heavy solvent and an organic light solvent, such as 30 per cent tributyl phosphate diluted with a paraffin hydrocarbon.

Scheibel [12–14] has described development work and process innovations for the separation of the rosin acids from the fatty acids in tall oil by fractional liquid–liquid extraction. He employed heptane as the light solvent and a mixture of methyl Cellosolve and water as the heavy solvent. Technical feasibility was demonstrated, but apparently the process was not commercialized, presumably because the cost of separation was too high. This is a typical example of a mixture of nonvolatile compounds not stable above about 500 °F. Distillation requires high-vacuum operation to keep the temperature down, and therefore separation by distillation is very costly. In such cases, fractional liquid–liquid extraction is often the most attractive alternative method of separation. On the other hand, this process is usually costly also, and it is perhaps not surprising to find commercial examples of fractional liquid–liquid extraction largely limited to the processing of quite valuable materials, such as the nuclear fuels referred to earlier.

3-7. General mathematical description

A mathematical description of the behavior of the fractional-extraction cascade in Fig. 3-11 will now be derived. It will be assumed that the solvent content of a solute feed stream is negligible; thus the flow rate of heavy solvent and of light solvent from stage to stage is assumed to be the same for all stages in the cascade (this also requires the assumption that the heavy and light solvents have no appreciable mutual solubility). Such a cascade may have more than one solute feed stream, and for the sake of generality Fig. 3-11 shows two such feed streams, one entering stage F and another entering stage G. The feed stream entering stage F may be introduced into the heavy-solvent flow from stage $F+1$ to stage F, or it may be introduced into the light-solvent flow from stage $F-1$ to stage F, or it may simply be introduced into the mixing chamber of stage F directly; if ideal stages are assumed, the net effect is the same in any case. The analysis will include the possibility that the light- and heavy-solvent feed streams may contain the solute components.

Attention is focused upon any given solute component. The material balances around the bottom stages are given by Eqs. (3-8) through (3-11) and can be transformed to the form of Eqs. (3-13) through (3-16). Equation (3-16) applies up to $j = F - 1$. The material balance for stage F is written

$$HX_{F+1} + LY_{F-1} + \mathscr{F} = HX_F + LY_F \tag{3-28}$$

where $\mathscr{F}$ represents the molal flow rate of the solute component in question in the solute feed stream entering stage F. This equation can be rearranged to the form

$$X_{F+1} - X_F = E(X_F - X_{F-1}) - \frac{\mathscr{F}}{H} \tag{3-29}$$

The material-balance equations for the stages between stage F and stage G take the form of Eqs. (3-11) and (3-16) for $j = F + 1 \cdots G - 1$. The material balance around stage G can be written as

$$X_{G+1} - X_G = E(X_G - X_{G-1}) - \frac{\mathscr{G}}{H} \tag{3-30}$$

where $\mathscr{G}$ represents the molal flow rate of the component in question in the solute feed stream entering stage G. For the stages above stage G, the material-balance relationship again takes the form of Eqs. (3-11) and (3-16) for $j = G + 1 \cdots N$.

An efficient method of reducing this system of simultaneous equations follows along the lines of the derivation of Eq. (2-53). The material balance for the first stage is Eq. (3-13).

$$X_2 - X_1 = E(X_1 - X_0) \tag{3-13}$$

Substitution of this equation into the right-hand side of Eq. (3-14), the material balance for the second stage, yields

$$X_3 - X_2 = E^2(X_1 - X_0) \tag{3-31}$$

Inserting this equation into the right-hand side of Eq. (3-15), the material balance for the third stage, yields

$$X_4 - X_3 = E^3(X_1 - X_0) \tag{3-32}$$

This general type of equation applies up to the material balance around stage $F - 1$,

$$X_F - X_{F-1} = E^{F-1}(X_1 - X_0) \tag{3-33}$$

Inserting Eq. (3-33) into the right-hand side of Eq. (3-29) yields

$$X_{F+1} - X_F = E^F(X_1 - X_0) - \frac{\mathscr{F}}{H} \tag{3-34}$$

The material balance around stage $F + 1$ again takes the form of Eq. (3-16); inserting Eq. (3-34) into the right-hand side of that equation yields

$$X_{F+2} - X_{F+1} = E^{F+1}(X_1 - X_0) - E\frac{\mathscr{F}}{H} \tag{3-35}$$

Inserting this equation into the right-hand side of the material balance around stage $F + 2$, which is of the form of Eq. (3-16), yields

$$X_{F+3} - X_{F+2} = E^{F+2}(X_1 - X_0) - E^2\frac{\mathscr{F}}{H} \tag{3-36}$$

Subsequent equations take this form up to and including the material balance around stage $G - 1$,

$$X_G - X_{G-1} = E^{G-1}(X_1 - X_0) - E^{G-F-1}\frac{\mathscr{F}}{H} \tag{3-37}$$

Inserting this equation into the right-hand side of Eq. (3-30), the material balance around stage G, yields

$$X_{G+1} - X_G = E^G(X_1 - X_0) - E^{G-F}\frac{\mathscr{F}}{H} - \frac{\mathscr{G}}{H} \tag{3-38}$$

The material balance around stage $G + 1$ again takes the form of Eq. (3-16); inserting Eq. (3-38) into that equation yields

$$X_{G+2} - X_{G+1} = E^{G+1}(X_1 - X_0) - E^{G-F+1}\frac{\mathscr{F}}{H} - E\frac{\mathscr{G}}{H} \tag{3-39}$$

Inserting this equation into the right-hand side of the material balance around stage $G + 2$, which is of the form of Eq. (3-16), yields

$$X_{G+3} - X_{G+2} = E^{G+2}(X_1 - X_0) - E^{G-F+2}\frac{\mathscr{F}}{H} - E^2\frac{\mathscr{G}}{H} \tag{3-40}$$

Subsequent equations take this form up to and including the material balance around stage N. For the present purpose, however, the series will be terminated with the material balance around stage $N - 1$,

$$X_N - X_{N-1} = E^{N-1}(X_1 - X_0) - E^{N-F-1}\frac{\mathscr{F}}{H} - E^{N-G-1}\frac{\mathscr{G}}{H} \tag{3-41}$$

Adding Eqs. (3-13), (3-31), . . ., (3-41) yields

$$\begin{aligned} X_N - X_1 = {} & (E + E^2 + E^3 + \cdots + E^{N-1})(X_1 - X_0) \\ & - (1 + E + E^2 + \cdots + E^{N-F-1})\frac{\mathscr{F}}{H} \\ & - (1 + E + E^2 + \cdots + E^{N-G-1})\frac{\mathscr{G}}{H} \end{aligned} \tag{3-42}$$

Substituting Y_N/D for X_N, multiplying through by H, and rearranging, Eq. (3-42) can be written as

$$\begin{aligned} \frac{H}{LD}LY_N = {} & (1 + E + E^2 + \cdots + E^{N-1})HX_1 \\ & - (E + E^2 + \cdots + E^{N-1})HX_0 \\ & - (1 + E + E^2 + \cdots + E^{N-F-1})\mathscr{F} \\ & - (1 + E + E^2 + \cdots + E^{N-G-1})\mathscr{G} \end{aligned} \tag{3-43}$$

Finally, using Eq. (2-56) and Eq. (3-17), Eq. (3-43) may be written as

$$\frac{LY_N}{E} = \frac{1 - E^N}{1 - E}HX_1 - E\frac{1 - E^{N-1}}{1 - E}HX_0 - \frac{1 - E^{N-F}}{1 - E}\mathscr{F} - \frac{1 - E^{N-G}}{1 - E}\mathscr{G} \tag{3-44}$$

The quantity LY_N is the molal flow rate of the solute component in the light-solvent stream leaving the top of the cascade from stage N; it will be given the symbol $\mathscr{T}$:

$$\mathscr{T} \equiv LY_N \tag{3-45}$$

The quantity HX_1 is the molal flow rate of the solute component in question in the heavy-solvent stream leaving the bottom of the cascade from stage 1; it will be given the symbol $\mathscr{B}$:

$$\mathscr{B} \equiv HX_1 \tag{3-46}$$

The quantity LY_0 is the flow rate of the solute component into the cascade with the light-solvent feed stream entering stage 1; it will be given the symbol $\mathscr{L}$:

$$\mathscr{L} \equiv LY_0 = EHX_0 \tag{3-47}$$

The second half of Eq. (3-47) follows from Eqs. (3-17) and (3-18). The flow rate of the solute component into the cascade with the heavy-solvent feed to stage N is represented by $\mathscr{H}$:

$$\mathscr{H} \equiv HX_{N+1} \tag{3-48}$$

With these definitions, Eq. (3-44) may be written as

$$\frac{1 - E}{E}\mathscr{T} = (1 - E^N)\mathscr{B} - (1 - E^{N-1})\mathscr{L} - (1 - E^{N-F})\mathscr{F} - (1 - E^{N-G})\mathscr{G} \tag{3-49}$$

The overall material balance around the entire cascade for the solute component in question may be written as

$$\mathscr{T} + \mathscr{B} = \mathscr{F} + \mathscr{G} + \mathscr{L} + \mathscr{H} \tag{3-50}$$

For given values of the extraction factor, E, the stage numbers, F, G, and N, and the input rates of a given component, $\mathscr{F}$, $\mathscr{G}$, $\mathscr{L}$ and $\mathscr{H}$, the simultaneous solution of Eqs. (3-49) and (3-50) yields values for $\mathscr{T}$ and $\mathscr{B}$ and thus for the percentage of that solute component that leaves the top of the cascade and the percentage that leaves from the bottom of the cascade. Eliminating $\mathscr{B}$ between Eqs. (3-49) and (3-50) yields, after some rearrangement,

$$\mathscr{T}\left(\frac{1}{E} - E^N\right) = \mathscr{F}(E^{N-F} - E^N) + \mathscr{G}(E^{N-G} - E^N) + \mathscr{L}(E^{N-1} - E^N) + \mathscr{H}(1 - E^N) \tag{3-51}$$

This equation gives the molal flow rate of a given solute component in the light-solvent stream leaving the top stage of the cascade as a function of the feed rates of that component in the solute feed streams and in the solvent feed streams. The coefficient of each feed rate on the right-hand side of Eq. (3-51) contains as its first term the extraction factor raised to a power that represents the number of stages *below* the *top* stage at which that particular feed stream entered the cascade; this equation could obviously be generalized to include any number of solute feed streams.

Eliminating $\mathscr{T}$ between Eqs. (3-50) and (3-51) yields, after some rearrangement,

$$\mathscr{B}\left(E - \frac{1}{E^N}\right) = \mathscr{F}\left(\frac{1}{E^{F-1}} - \frac{1}{E^N}\right) + \mathscr{G}\left(\frac{1}{E^{G-1}} - \frac{1}{E^N}\right) + \mathscr{L}\left(1 - \frac{1}{E^N}\right) + \mathscr{H}\left(\frac{1}{E^{N-1}} - \frac{1}{E^N}\right) \tag{3-52}$$

This equation gives the molal flow rate of a given component in the heavy-solvent stream leaving the bottom stage of the cascade as a function of the various feed rates of that component. Equation (3-52) is symmetrical with Eq. (3-51). The extraction factor E in Eq. (3-51) has been replaced by its reciprocal in Eq. (3-52). The coefficient of each of the feed rates in Eq. (3-52) contains the reciprocal of the extraction factor raised to a power equal to the number of stages *above* the *first* stage at which that particular feed stream is fed to the cascade. Equation (3-52) can obviously be generalized to include any number of solute feed streams.

It must be emphasized that Eqs. (3-49) through (3-52) may be written for each solute component in the cascade. In this linearized analysis it is apparent that the presence of one solute component will not affect the way another solute component distributes between the top and bottom product streams, but since different solute components will generally have different E values, they will generally distribute in different ways.

3-8. Fractional extraction with one solute feed stream and with pure solvent feed streams

Consider a fractional-extraction cascade with solvent feed streams which are solute free and with a single solute feed stream into stage F. In this case, Eqs. (3-51) and (3-52) simplify to

$$\mathscr{T}\left(\frac{1}{E} - E^N\right) = \mathscr{F}(E^{N-F} - E^N) \tag{3-53}$$

$$\mathscr{B}\left(E - \frac{1}{E^N}\right) = \mathscr{F}\left(\frac{1}{E^{F-1}} - \frac{1}{E^N}\right) \tag{3-54}$$

These equations give directly the fraction of a solute component that leaves the top of the cascade and the fraction that leaves the bottom of the cascade in terms of the extraction factor for that component, the number of stages in the cascade, and the feed-stage number. Another useful equation can be obtained by dividing Eq. (3-53) by Eq. (3-54) to give, after rearrangement,

$$\frac{\mathscr{T}}{\mathscr{B}} = E^{N-F+1}\frac{E^F - 1}{E^{N-F+1} - 1} \tag{3-55}$$

The feed flow rate, $\mathscr{F}$, has been canceled from this equation, and the exit-flow-rate distribution has been represented by the ratio, $\mathscr{T}/\mathscr{B}$, of the exit flow from the top of the cascade to the exit flow from the bottom of the cascade.

For the special case of a cascade with the solute feed stream entering into the middle stage, $N - F = F - 1$, and Eq. (3-55) simplifies to

$$\frac{\mathscr{T}}{\mathscr{B}} = E^F = E^{(N+1)/2} \qquad \left(\text{for } F = \frac{N+1}{2}\right) \tag{3-56}$$

Equations (3-51) through (3-56) apply separately to each solute component in the cascade. Consider a cascade with two solute components, component A and component B, with distribution coefficients D_A and D_B, respectively, between the heavy-solvent and the light-solvent phases. Assume that the cascade has the solute feed into the middle stage such that Eq. (3-56) applies. For component A, Eq. (3-56) gives

$$\frac{\mathscr{T}_A}{\mathscr{B}_A} = E_A^F \tag{3-57}$$

Similarly, for component B,

$$\frac{\mathscr{T}_B}{\mathscr{B}_B} = E_B^F \tag{3-58}$$

Dividing Eq. (3-57) by Eq. (3-58) gives

$$\frac{\mathscr{T}_A/\mathscr{B}_A}{\mathscr{T}_B/\mathscr{B}_B} = \left(\frac{E_A}{E_B}\right)^F = \left(\frac{D_A}{D_B}\right)^F \equiv \alpha^F = \alpha^{(N+1)/2} \qquad \left(\text{for } F = \frac{N+1}{2}\right) \tag{3-59}$$

where it is assumed that D_A is greater than D_B, and thus that the relative distribution coefficient, α, is defined as D_A/D_B. The left-hand side of Eq. (3-59) is an excellent measure of the degree to which the cascade can separate component A from component B. If most of component A leaves the cascade in the light-solvent stream from the top stage, $\mathscr{T}_A/\mathscr{B}_A$ will be a large number. If most of component B leaves the cascade with the heavy-solvent stream leaving the bottom of the cascade, $\mathscr{T}_B/\mathscr{B}_B$ will be a very small number. If both of these are true, the left-hand side of Eq. (3-59) will be a very large number, indicating that a high degree of separation between the components has been achieved. Actually, the ratio shown on the left-hand side of Eq. (3-59) is not a unique measure of the degree of separation achieved in the cascade but is a good measure of the degree of separation which can be achieved with proper adjustment of the solvent flow rates. This ratio is called the *separating power* of the cascade and is given the symbol S:

$$S_{A,B} \equiv \frac{\mathscr{T}_A/\mathscr{B}_A}{\mathscr{T}_B/\mathscr{B}_B} \tag{3-60}$$

Equation (3-59) reveals that, for the particular cascade under consideration, $S_{A,B}$ is a unique function of the relative distribution coefficient, α, and the number of stages in the cascade, and is independent of the flow rates L and H. But the nature of the separation achieved between components A and B is very much dependent upon the solvent flow-rate ratio, as will be shown in the following section.

3-9. Symmetrical split between solutes

Consider, for example, a solute feed that is an equimolal mixture of components A and B. Suppose that it is desired to produce a light-solvent product at the top of the cascade which contains 99 per cent of the quantity of component A fed to the system and 1 per cent of the quantity of component B fed to the system, and to produce a heavy-solvent product from the bottom of the cascade which contains 1 per cent of the quantity of component A fed to the system and 99 per cent of the quantity of component B fed to the system. In this case, $\mathscr{T}_A/\mathscr{B}_A = 99$, and $\mathscr{T}_B/\mathscr{B}_B = \frac{1}{99}$. Thus $S_{A,B} = 99^2$, and the number of stages in the cascade must be chosen such that $\alpha^F = 99^2$. But Eq. (3-57) also requires that E_A^F must equal 99, and Eq. (3-58) requires that E_B^F must equal $\frac{1}{99}$. Thus

$$E_A^F \equiv \left(D_A \frac{L}{H}\right)^F = 99 \tag{3-61}$$

$$E_B^F \equiv \left(D_B \frac{L}{H}\right)^F = \frac{1}{99} \tag{3-62}$$

Multiplying Eq. (3-61) by Eq. (3-62) gives

$$\left[D_A D_B \left(\frac{L}{H}\right)^2\right]^F = 1 \tag{3-63}$$

from which it follows that

$$\frac{H}{L} = \sqrt{D_A D_B} \tag{3-64}$$

which is necessary for a symmetrical split, $\mathscr{T}_A/\mathscr{B}_A = \mathscr{B}_B/\mathscr{T}_B$, in a cascade fed at the middle stage. Thus in this symmetrical case the flow-rate ratio should be chosen such that H/L is the geometric mean of the distribution coefficients for components A and B. With this choice it follows that

$$E_A = \sqrt{\frac{D_A}{D_B}} = \sqrt{\alpha} \tag{3-65}$$

$$E_B = \sqrt{\frac{D_B}{D_A}} = \frac{1}{\sqrt{\alpha}} \tag{3-66}$$

The ratio E_A/E_B is always equal to α; this particular choice of the flow-rate ratio makes E_A greater than unity and E_B less than unity such that the product $E_A E_B$ is equal to unity. Component A, with $E_A > 1$, is washed to the top of the cascade, and component B, with $E_B < 1$, is washed to the bottom of the cascade.

If the flow-rate ratio is chosen according to Eq. (3-64) and the number of stages is chosen such that $\alpha^F = 99^2$, the desired separation will be achieved. On the other hand, even if the number of stages is chosen such that α^F is equal to 99^2, varying the flow-rate ratio from that given by Eq. (3-64) will destroy the symmetrical separation between components A and B. For example, if instead of choosing H/L according to Eq. (3-64), this ratio were chosen smaller by a factor $\sqrt{\alpha}$, E_A would be equal to α and E_B would be equal to unity. In this case, according to Eqs. (3-57) and (3-58), $\mathcal{T}_A/\mathcal{B}_A$ would be equal to 99^2 but $\mathcal{T}_B/\mathcal{B}_B$ would be equal to unity. Thus, by choosing L/H too large by the factor $\sqrt{\alpha}$, a far better job of keeping component A out of the bottom of the cascade has been done, but at the expense of allowing half of component B to leave the top of the cascade while the other half leaves the bottom. This would produce a very pure component B at the bottom of the cascade but would contaminate the top product with half of the amount of species B fed to the system. Conversely, if H/L were chosen larger than Eq. (3-64) dictates by the factor $\sqrt{\alpha}$, $\mathcal{T}_B/\mathcal{B}_B$ would be equal to $1/99^2$, but $\mathcal{T}_A/\mathcal{B}_A$ would equal unity. Thus the product at the top of the column would have a very low concentration of species B, but the bottom product would be contaminated with half of the amount of species A fed to the system. All these cases correspond to $S_{A,B} = 99^2$, but only if the flow-rate ratio is chosen according to Eq. (3-64) will the split between the two solutes be symmetrical, and thus relatively pure species A be obtained at the top of the cascade and relatively pure species B be obtained at the bottom.

EXAMPLE PROBLEM 3-7 A mixture containing 50 wt % linoleic acid and 50 wt % abietic acid is to be separated in a fractional-extraction cascade into a top product containing 98 per cent of the linoleic acid and 2 per cent of the abietic acid in the feed stream and a bottom product containing 2 per cent of the linoleic acid and 98 per cent of the abietic acid in the feed stream. The light solvent will be heptane and the heavy solvent will be methyl Cellosolve containing 10 per cent water by volume; thus, with X and Y representing mass ratios, the distribution coefficients will be 2.17 and 1.57 for linoleic acid and abietic acid, respectively. If the cascade is to have the solute feed stream entering at the middle stage, how many stages are required to achieve the desired separation? What should be the ratio of the light-solvent feed rate to the heavy-solvent feed rate?

Solution: Employing the subscripts A for linoleic acid, the solute with the greater distribution coefficient, and B for abietic acid, Eq. (3-59) requires that

$$\frac{98/2}{2/98} = \left(\frac{2.17}{1.57}\right)^F$$

for which the solution is

$$F = 24$$

Thus 47 stages are required in the cascade with the solute feed at stage 24. The ratio of the solvent flow rates is given by Eq. (3-64) as

$$\frac{H}{L} = \sqrt{(2.17)(1.57)} = 1.847 \ \frac{\text{pounds of heavy solvent}}{\text{pound of light solvent}}$$

EXAMPLE PROBLEM 3-8 Referring to the cascade design of Example Problem 3-7, if the heavy-solvent to light-solvent flow-rate ratio is increased 5 per cent, what will be the linoleic and abietic acid contents of the top and bottom product streams?

Solution: A 5 per cent increase in H/L would result in

$$\frac{H}{L} = (1.05)(1.847) = 1.939$$

The extraction factors would then become

$$E_A = \frac{2.17}{1.939} = 1.12$$

$$E_B = \frac{1.57}{1.939} = 0.81$$

Equations (3-57) and (3-58) yield

$$\frac{\mathcal{T}_A}{\mathcal{B}_A} = 1.12^{24} = 15.2$$

$$\frac{\mathcal{T}_B}{\mathcal{B}_B} = 0.81^{24} = 0.00635$$

Note that the ratio of these two values is 49^2, as in Example Problem 3-7. On the basis of 1 lb of linoleic acid and 1 lb of abietic acid fed in the solute feed stream, the acid contents of the top product are

$$\mathcal{T}_A = \frac{15.2}{16.2} = 0.9383$$

$$\mathcal{T}_B = \frac{0.00635}{1.00635} = 0.00631$$

and the acid contents of the bottom product are

$$\mathscr{B}_A = \frac{1}{16.2} = 0.0617$$

$$\mathscr{B}_B = \frac{1}{1.00635} = 0.9937$$

The denominators used above are chosen such that for each acid, $\mathscr{T}$ and $\mathscr{B}$ will add to 1 lb and will be in the ratios computed earlier. The acid content of the top product stream is approximately 99.3 per cent linoleic acid and only 0.7 per cent abietic acid. But the acid content of the bottom product stream is approximately 5.8 per cent linoleic acid and only 94.2 per cent abietic acid. The excessive heavy-solvent flow rate has resulted in doing a better job of removing the abietic acid from the top product but has resulted in too much contamination of the bottom product with linoleic acid.

EXAMPLE PROBLEM 3-9 A mixture containing 25 wt % oleic acid, 40 wt % linoleic acid, and 35 wt % abietic acid is fed at a steady rate of 1000 lb/hr to a fractional extraction cascade. The light and heavy solvents are those of Example Problem 3-7; therefore, the distribution coefficients, expressed in terms of mass ratios, are equal to 4.14, 2.17, and 1.57 for oleic, linoleic, and abietic acids, respectively. The cascade contains the equivalent of 41 theoretical stages, with the feed being added to stage 21.

(a) What should be the solvent flow-rate ratio for a symmetrical split between linoleic and abietic acids? What will be the resulting flow rates of the acids in the top and bottom product streams?

(b) Repeat for a symmetrical split between oleic and linoleic acids.

Solution

(a) Using Eq. (3-64),

$$\frac{H}{L} = \sqrt{(2.17)(1.57)} = 1.847 \text{ lb of heavy solvent/lb of light solvent}$$

The extraction factors become

$$E_A = \frac{4.14}{1.847} = 2.24$$

$$E_B = \frac{2.17}{1.847} = 1.176$$

$$E_C = \frac{1.57}{1.847} = 0.85$$

where A, B, and C represent oleic, linoleic, and abietic acids, respectively. Equation (3-56) then gives

$$\frac{\mathscr{T}_A}{\mathscr{B}_A} = (2.24)^{21} = 23 \times 10^6$$

$$\frac{\mathscr{T}_B}{\mathscr{B}_B} = (1.176)^{21} = 30$$

$$\frac{\mathscr{T}_C}{\mathscr{B}_C} = (0.85)^{21} = \frac{1}{30}$$

Employing these ratios, the material balances are readily calculated, as shown in Table 3-1. The calculation of the quantities tabulated will be illustrated for linoleic

Table 3-1

		Flow Rate (lb/hr) in	
Acid	*Feed*	*Top Product*	*Bottom Product*
Oleic	250	249.99999	$\sim 10^{-5}$
Linoleic	400	387.1	12.9
Abietic	350	11.3	338.7
Total	1000	648.4	351.6

acid. The overall material balance is

$$\mathscr{T}_B + \mathscr{B}_B = 400$$

Rearranging gives

$$\mathscr{B}_B\left(\frac{\mathscr{T}_B}{\mathscr{B}_B} + 1\right) = 400$$

or

$$\mathscr{B}_B(30 + 1) = 400$$

$$\mathscr{B}_B = \frac{400}{31} = 12.9$$

$$\mathscr{T}_B = 400 - 12.9 = 387.1$$

(b) Equation (3-64) becomes

$$\frac{H}{L} = \sqrt{(4.14)(2.17)} = 2.997$$

The extraction factors become

$$E_A = \frac{4.14}{2.997} = 1.38$$

$$E_B = \frac{2.17}{2.997} = 0.725$$

$$E_C = \frac{1.57}{2.997} = 0.525$$

Equation (3-56) then gives

$$\frac{\mathscr{T}_A}{\mathscr{B}_A} = (1.38)^{21} = 860$$

$$\frac{\mathscr{T}_B}{\mathscr{B}_B} = (0.725)^{21} = \frac{1}{860} = 0.00117$$

$$\frac{\mathscr{T}_C}{\mathscr{B}_C} = (0.525)^{21} = 1.3 \times 10^{-6}$$

The calculated product rates are shown in Table 3-2.

Table 3-2

	Flow Rate (lb/hr) in		
Acid	*Feed*	*Top Product*	*Bottom Product*
Oleic	250	249.71	0.29
Linoleic	400	0.465	399.535
Abietic	350	$\sim 5 \times 10^{-4}$	349.9995
Total	1000	250.1755	749.8245

3-10. Solute buildup in the middle of the cascade

For simplicity, attention will be focused upon a cascade with a single solute feed into the middle stage. Equation (3-59) determines the number of stages required for a given degree of separation, and Eq. (3-64) determines the solvent flow-rate ratio required for a symmetrical split. But none of these equations indicates how large the flow rate of heavy solvent and the flow rate of light solvent must be; only their ratio is determined by Eq. (3-64), and only their ratio enters into any of the equations so far derived. Indeed, if the several assumptions made to justify a *linear* model of the cascade were unconditionally correct, the values of H and L would have no effect upon the performance of the system; only the ratio L/H would be important. But, for a given solute feed rate, small values of H and L will obviously result in high solute concentrations in the various streams within the cascade, while large values of H and L will result in low solute concentrations. As explained earlier, the assumption that the distribution coefficient for each solute is constant, independent of the concentration level of the solute and of the presence of other solutes, will surely break down at sufficiently high solute concentrations. Furthermore, the assumption that the light- and heavy-solvent components do not dissolve in each other to an appreciable extent will generally break down at high solute concentrations and, indeed, it is often found that at sufficiently high concentrations of the solutes the two solvent components are completely soluble in each other, and only one liquid phase exists. Under such conditions, the liquid–liquid extraction system could not operate. Even before this happens, as the solute content of the solvent phases is increased, the distribution coefficients of the various solute components will generally approach each other such that the relative distribution coefficient, α, will approach unity, and separation becomes increasingly difficult. Furthermore, the densities of the two phases will generally approach each other at high solute concentrations, and a point may be reached where the physical separation of the two liquid phases by gravity or by centrifugal fields will become prohibitively difficult.

So the assumptions that make possible a linear model of the cascade become less and less accurate and the separation becomes more and more difficult at higher and higher solute concentrations. Accordingly, for a fixed feed rate of the solute components, the values of the heavy- and light-solvent feed rates must be sufficiently high so as to maintain the solute concentrations within the cascade sufficiently low that a separation is possible by liquid–liquid extraction. It should not be assumed that the solvent flow rates should be maintained sufficiently high and thus the concentrations of the solutes sufficiently low that the assumptions which lead to a linear model are valid; this would make the cascade easy to design, but it would probably not, in general, lead to the economically optimum cascade design. High solvent flow rates require large contacting equipment and result in high costs of separating the solvent from the solute component in the product stream. On the other hand, low solvent flow rates result in higher solute concentrations, which generally result in poor values of the relative distribution coefficient and, therefore, an increased number of stages in the cascade. If the densities of the two solvent phases closely approach each other, this could also result in increased size of the settling equipment. Thus it is obvious that there is an optimum choice for the magnitudes of the heavy- and light-solvent feed rates which will minimize the cost of the separation process. It is clear that this economic balance involves the nonlinearities that have been ignored in the linear treatment presented here. It is beyond the scope of this presentation to treat these effects quantitatively; rather, the linearized mathematical model developed here will be used as an approximation, and an arbitrary upper limit on the concentrations of the solute components in the solvent phases will be imposed to account, in a very approximate way, for the various considerations mentioned above.

In this connection it is extremely important to realize that the concentrations of the solute components build up very substantially toward the middle of the cascade. If Eq. (3-13) and Eqs. (3-31) through (3-33) are added together, the result is

$$X_F - X_1 = (E + E^2 + \cdots + E^{F-1})(X_1 - X_0) \tag{3-67}$$

For pure solvent feed streams, $X_0 = 0$, and, using Eq. (2-56), this equation becomes

$$\frac{X_F}{X_1} = \frac{HX_F}{\mathscr{B}} = \frac{1 - E^F}{1 - E} \tag{3-68}$$

For a cascade with one solute feed stream entering at stage F, it is clear from the symmetry of the cascade that the analogous equation for the top of the cascade is

$$\frac{Y_F}{Y_N} = \frac{LY_F}{\mathscr{T}} = \frac{1 - 1/E^{N-F+1}}{1 - 1/E} \tag{3-69}$$

When the relative distribution coefficient between two components being separated is near unity, the extraction factors for the two components will generally be near unity, and Eqs. (3-68) and (3-69) will indicate a large buildup of the solute components at the middle of the cascade.

EXAMPLE PROBLEM 3-10 For the cascade design of Example Problem 3-7, what should be the light- and heavy-solvent flow rates if the feed rate of the mixture of linoleic and abietic acids is 1000 lb/hr and if it is stipulated that the maximum acid concentration of either solvent phase at any point within the cascade should not exceed 0.08 lb of total acids per pound of solvent?

Solution: For linoleic acid, 500 lb/hr is fed to the system and 490 lb/hr leaves with the light-solvent product stream at the top, while 10 lb/hr leaves with the heavy-solvent product stream at the bottom. Thus

$$X_{1_A} = \frac{10}{H}$$

$$Y_{N_A} = \frac{490}{L}$$

Similarly, for abietic acid,

$$X_{1_B} = \frac{490}{H}$$

$$Y_{N_B} = \frac{10}{L}$$

From Eqs. (3-65) and (3-66),

$$E_A = \frac{1}{E_B} = \sqrt{\frac{2.17}{1.57}} = 1.176$$

Using Eq. (3-68) for linoleic acid,

$$\frac{X_{F_A}}{X_{1_A}} = \frac{(1.176)^{24} - 1}{1.176 - 1} = 273$$

Similarly, for abietic acid,

$$\frac{X_{F_B}}{X_{1_B}} = \frac{1 - (1/1.176)^{24}}{1 - 1/1.176} = 6.56$$

The total acid content of the heavy-solvent stream leaving the feed stage is

$$X_{F_A} + X_{F_B} = 273\frac{10}{H} + 6.56\frac{490}{H} = \frac{5940}{H}$$

In order that the total acid content of this heavy-solvent stream be less than 0.08 lb of total acid per pound of heavy solvent, the heavy-solvent flow rate must be at least as high as

$$H = \frac{5940}{0.08} = 74{,}300 \text{ lb/hr}$$

Since $H/L = 1.847$, the corresponding value of the light-solvent flow rate would be

$$L = \frac{74{,}300}{1.847} = 40{,}200 \text{ lb/hr}$$

These solvent flow rates will assure that the total acid content of the heavy-solvent stream leaving the feed stage shall not exceed 0.08 lb of total acid per pound of solvent.

Next, a check should be made upon the light-solvent stream leaving the feed stage. Since for each component

$$Y_F = DX_F$$

it follows that

$$Y_{F_A} + Y_{F_B} = (2.17)(273)\frac{10}{H} + (1.57)(6.56)\frac{490}{H}$$

$$= \frac{10{,}970}{H}$$

To keep the total acid content of the light solvent less than 0.08 lb of total acids per pound of light solvent, the heavy-solvent flow rate must be at least as great as

$$H = \frac{10{,}970}{0.08} = 137{,}000 \text{ lb/hr}$$

with a corresponding light-solvent flow rate of

$$L = \frac{137{,}000}{1.847} = 74{,}200 \text{ lb/hr}$$

These flow rates are greater than those required to keep $X_{F_A} + X_{F_B}$ less than 0.08 (this must be so when both D values exceed unity); therefore, these later values represent the required solvent flow rates.

There is no need to check the streams entering the feed stage because their solute concentrations will be less than those of the streams leaving the feed stage for any cascade with a single feed stage and with pure solvent feed streams. To demonstrate that this is so, consider the bottom $F - 1$ stages as a simple countercurrent cascade in which a pure light-solvent wash is contacted with a rich heavy-solvent stream from stage F. Clearly, X_F will be greater than X_{F-1} for any component. Since, for any component, $Y_F = DX_F$ and $Y_{F-1} = DX_{F-1}$, it follows that Y_F is greater than Y_{F-1}. A similar argument leads to the conclusion that X_F is greater than X_{F+1}.

3-11. Graphical representation of fractional-extraction cascade

A fractional-extraction cascade can be represented by a McCabe–Thiele type of diagram such as those in Figs. 3-5 and 3-6, but the diagram contains two material-balance lines, one above the equilibrium line and one below it. Figure 3-12 shows such a diagram for a nine-stage cascade with solute feed

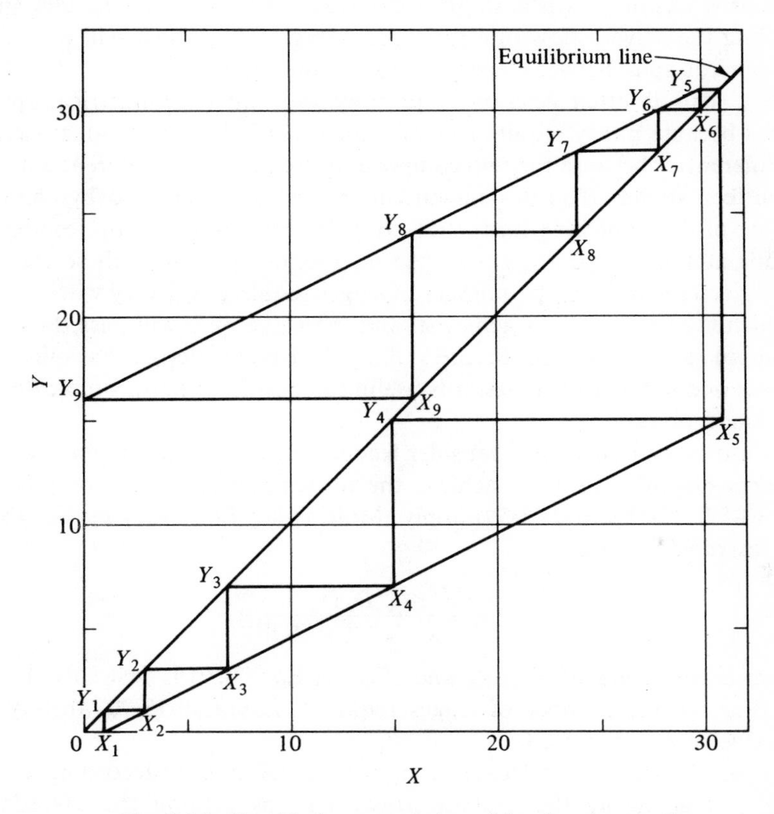

Fig. 3-12. Graphical representation of fractional extraction.

into the fifth stage. Figure 3-12 represents the concentrations of a solute with $D = 1$ for a solvent flow-rate ratio $H/L = \frac{1}{2}$; thus $E = 2$. For this case, $X_1 = 1$ and $Y_9 = 16$. With $H/L = \frac{1}{2}$, this corresponds to $\mathscr{T}/\mathscr{B} = 32$ for this component, as required by Eq. (3-59).

The lower material-balance line represents the bottom of the cascade, stages 1 through 4, and the upper material-balance line represents the top of the cascade, stages 6 through 9. Both lines have a slope of $\frac{1}{2}$, but one lies above the equilibrium line and the other lies below it. In the bottom of the cascade, X builds up from $X_1 = 1$ to $X_5 = 31$, as required by Eq. (3-68). Above the feed stage, Y falls from $Y_5 = 31$ to $Y_9 = 16$.

A separate diagram with a different slope of the equilibrium line but with the same slope for the material-balance lines is needed for each solute component.

3-12. Unsymmetrical cascades

For a cascade with a single feed stream added to the middle stage, choosing the solvent flow-rate ratio according to Eq. (3-64) will produce a symmetrical split between components A and B, that is, with $\mathscr{T}_A/\mathscr{B}_A = \mathscr{B}_B/\mathscr{T}_B$. This is often desirable, but sometimes an unsymmetrical split is desired because purity requirements for the top and bottom product streams are different or because the two components are present in different amounts in the feed stream. The unsymmetrical split may be achieved by choosing a value of H/L different from that given by Eq. (3-64), or by feeding the cascade at some stage away from the middle, or by both of these changes. For a given separation, the number of stages required will vary with the way in which the asymmetry is achieved, and the value of N will pass through a minimum as H/L is varied. Before seeking the cascade design that minimizes the required number of stages, it is useful to consider the two simple designs mentioned above.

Case 1. As a first case, consider feeding the cascade at the middle stage but choosing H/L so as to achieve the unsymmetrical split. For this case Eqs. (3-57), (3-58), and (3-59) apply. Multiplying Eq. (3-57) by Eq. (3-58) and rearranging yields

$$\frac{H}{L} = \frac{\sqrt{D_A D_B}}{[(\mathscr{T}_A/\mathscr{B}_A)(\mathscr{T}_B/\mathscr{B}_B)]^{1/2F}} \tag{3-70}$$

For specified values of $\mathscr{T}_A/\mathscr{B}_A$ and $\mathscr{T}_B/\mathscr{B}_B$, Eq. (3-59) is first solved for F and thus for the number of stages required. Equation (3-70) then yields the value of H/L required.

Case 2. As a second case, consider choosing H/L according to Eq. (3-64) but achieving the unsymmetrical split by feeding the cascade off center. Perhaps unexpectedly, this design requires the same total number of stages as does the design of case 1. This is demonstrated as follows. When H/L is chosen according to Eq. (3-64), Eqs. (3-65) and (3-66) apply, and thus the E values for the two components are reciprocals of each other. Since the cascade is fed off center, Eq. (3-55) is used for each component. Employing Eqs. (3-65) and (3-66), Eq. (3-55) takes the forms

$$\frac{\mathscr{T}_A}{\mathscr{B}_A} = \frac{\alpha^{(N-F+1)/2}(\alpha^{F/2} - 1)}{\alpha^{(N-F+1)/2} - 1} \tag{3-71}$$

$$\frac{\mathscr{T}_B}{\mathscr{B}_B} = \frac{\alpha^{-(N-F+1)/2}(\alpha^{-F/2} - 1)}{\alpha^{-(N-F+1)/2} - 1} \tag{3-72}$$

Dividing Eq. (3-71) by Eq. (3-72) yields, after considerable rearrangement,

$$\frac{\mathscr{T}_A/\mathscr{B}_A}{\mathscr{T}_B/\mathscr{B}_B} = \alpha^{(N+1)/2} \tag{3-73}$$

Comparing this equation with Eq. (3-59), it is clear that the number of stages required is identical to that of case 1, in which the feed was added to the middle stage.

Case 3. Since the required number of stages is the same for cases 1 and 2, it seems reasonable to assume that the minimum in the curve of N versus H/L will occur approximately half way between the H/L values for cases 1 and 2. As a useful rule, the geometric mean of the H/L values corresponding to cases 1 and 2 will be used as an approximation to the H/L value which will result in the minimum number of stages for a given separation:

$$\left(\frac{H}{L}\right)_{\text{case 3}} = \sqrt{\left(\frac{H}{L}\right)_{\text{case 1}}\left(\frac{H}{L}\right)_{\text{case 2}}} \tag{3-74}$$

Employing this value of H/L, Eq. (3-55) is written for component A and for component B and these two equations are solved simultaneously for N and F. Other values of H/L can also be chosen and the simultaneous solution repeated. The value of H/L given by Eq. (3-74) will usually be quite close to the value that minimizes N. Of course, minimizing N will not necessarily result in the optimum design, because the solute-buildup and thus the solvent-circulation rates will generally be different for these different cases. These procedures are illustrated in the following example problem.

EXAMPLE PROBLEM 3-11 A mixture contains 50, 5, and 45 wt % of oleic, linoleic, and abietic acids, respectively. It is to be separated in a fractional-extraction cascade into a top product containing linoleic acid and abietic acid in the weight ratio 99:1 and a bottom product containing linoleic acid and abietic acid in the weight ratio 1:99. The solvents will be the same as those in Example Problems 3-7 and 3-9; thus the distribution coefficients are 4.14, 2.17, and 1.57 for oleic, linoleic, and abietic acids, respectively. Determine the number of stages in the cascade, the feed-stage number, and the solvent flow-rate ratio for the cascade design that minimizes the number of stages required. For this design also determine the amount of oleic acid that will contaminate the bottom product.

Solution: The specifications on the product streams are written in terms of linoleic and abietic acids (which are therefore called *key components*); the oleic acid contamination of the bottom product is of interest but is not a specification. The flow rates of linoleic and abietic acids in the solute feed and the top and bottom product streams are summarized in Table 3-3, which is based upon a solute feed rate of 200

Table 3-3

Component	Mass Flow Rate (lb/unit time) in		
	Solute Feed	*Top Product*	*Bottom Product*
Linoleic acid	10	9.092	0.908
Abietic acid	90	0.092	89.908
Total	100	9.184	90.816

lb per unit time. These values are obtained readily from overall material balances for the components plus the specified ratios in the product streams. Denoting linoleic acid as component A and abietic acid as component B, it follows from Table 3-3 that

$$\frac{\mathcal{T}_A}{\mathcal{B}_A} = \frac{9.092}{0.908} = 10.01$$

$$\frac{\mathcal{T}_B}{\mathcal{B}_B} = \frac{0.092}{89.908} = \frac{1}{977}$$

Thus it is seen that the split is quite asymmetric because of the large ratio of component B to component A in the solute feed, even though the ratios of the key components in the top and bottom product streams are the same.

For case 1, in which the solute feed stream is fed to the middle stage, Eq. (3-59) applies:

$$\left(\frac{2.17}{1.57}\right)^F = (10.01)(977) = 9780$$

The solution is

$$F = 28.4$$

and with this value of F, N equals 55.8. Of course, F and N must be integers in a cascade with ideal or equilibrium stages, but in real cascades a stage efficiency is usually needed, and the consideration of noninteger numbers of ideal stages is expedient. The flow-rate ratio is given by Eq. (3-70):

$$\frac{H}{L} = \frac{\sqrt{(2.17)(1.57)}}{(10.01/977)^{1/56.8}} = \frac{1.847}{0.923} = 2.00$$

For case 2, the flow-rate ratio is chosen according to Eq. (3-64). Thus, as in Example Problem 3-7,

$$\frac{H}{L} = 1.847$$

For this case, also, $N = 55.8$, as given by Eq. (3-73).

There is no real need to determine the feed-stage number, F, for this case, but it can be done by using Eq. (3-55) written for component A or component B. Thus, for component A,

$$E_A = \frac{2.17}{1.847} = 1.176$$

and Eq. (3-55) becomes

$$10.01 = \frac{1.176^{56.8-F}(1.176^F - 1)}{1.176^{56.8-F} - 1}$$

Solving by trial and error gives $F = 14.8$; therefore, the feed point is way off center.

Case 3. For this case, H/L is chosen according to Eq. (3-74):

$$\frac{H}{L} = \sqrt{(2.00)(1.847)} = 1.922$$

The extraction factors become

$$E_A = \frac{2.17}{1.922} = 1.128$$

$$E_B = \frac{1.57}{1.922} = \frac{1}{1.226}$$

Equation (3-55) for component A becomes

$$10.01 = 1.128^{N-F+1}\frac{1.128^F - 1}{1.128^{N-F+1} - 1}$$

Assuming that 1.128^{N-F+1} is large compared with unity, this equation requires, approximately, that

$$1.128^F = 11.01$$

Thus

$$F = 19.9$$

Equation (3-55) for component B becomes

$$\frac{1}{977} = \frac{1}{1.226^{N-F+1}}\frac{1 - (1/1.226^F)}{1 - (1/1.226^{N-F+1})}$$

For an approximation, this may be reduced to

$$\frac{1}{977} \cong \frac{1}{1.226^{N-F+1}}\left(1 - \frac{1}{1.226^{19.9}}\right)$$

Therefore,

$$N - F + 1 \cong 33.7$$

Using this value,

$$1.128^{N-F+1} \cong 58$$

This value is large compared to unity, but a slightly better answer can be obtained by returning to the equation for component A.

$$10.01 = \frac{58}{57}(1.128^F - 1)$$

From this it follows that

$$F = 19.8$$

Finally, returning to the equation for component B with this value of F,

$$\frac{1}{977} = \frac{(1/1.226^{N-F+1})[1 - (1/1.226^{19.8})]}{1 - (1/1.226^{N-F+1})}$$

The solution remains

$$N - F + 1 = 33.7$$

Thus the computed result is $N = 52.5$, $F = 19.8$. Similar calculations at other values of H/L yield the results in Table 3-4, in which the results of cases 1 and 2 have also been included. As the value of H/L increases, the solute feed must be introduced

Table 3-4

H/L	*F*	*N*	*Remarks*
1.847	14.8	55.8	Case 2
1.885	17	53.6	
1.922	19.8	52.5	Case 3
1.962	23.4	53.2	
2.00	28.4	55.8	Case 1

higher and higher in the cascade. The total number of stages goes through a minimum, and it can be seen by plotting the values in the table that case 3 is quite close to that minimum point.

Continuing with the case 3 design, denoting oleic acid as component C,

$$E_C = \frac{4.14}{1.922} = 2.15$$

Equation (3-55) for oleic acid then takes the form

$$\frac{\mathcal{T}_C}{\mathcal{B}_C} = \frac{(2.15)^{33.7}[(2.15)^{19.8} - 1]}{(2.15)^{33.7} - 1}$$

$$= 3.8 \times 10^6$$

Therefore, on the basis of 100 lb per unit time for the feed rate of oleic acid, 2.6×10^{-5} lb appears in the bottom-product stream and the rest appears in the top-product stream. The material-balance table, Table 3-5, shows the way the three acids distribute between the top and bottom product streams. The table is based upon a feed rate of 200 lb of total acids per unit time.

Table 3-5

	Mass Flow Rate (lb/unit time) in		
Component	*Solute Feed*	*Top Product*	*Bottom Product*
Oleic acid	100	$100 - 2.6 \times 10^{-5}$	2.6×10^{-5}
Linoleic acid	10	9.092	0.908
Abietic acid	90	0.092	89.908
Total	200	109.184	90.816

It should be realized that the amount of oleic acid in the bottom product will be quite different for the various designs corresponding to different value of H/L. Since E_C^{N-F+1} is such a large number, Eq. (3-55) for oleic acid can be approximated by

$$\frac{\mathscr{T}_C}{\mathscr{B}_C} \cong E_C^F - 1$$

For the design of case 2, $\mathscr{T}_C/\mathscr{B}_C$ equals 1.6×10^4 because F is relatively small. For the design of case 1, E_C is smaller, but F is so much larger that $\mathscr{T}_C/\mathscr{B}_C$ equals 9×10^8. Therefore, the various designs correspond to different distributions for the oleic acid as well as to different numbers of stages.

The solute buildup at the feed stage is also different for the different designs. Using Eq. (3-69) with the values of $N - F + 1$, E, and $\mathscr{T}$ for each acid for each case, on the basis of a feed rate of 200 lb of feed per unit time, gives the results shown in Table 3-6. While case 1 requires more stages than case 3, the solvent requirements

Table 3-6

	LY_F, *lb/unit time*		
Acid	*Case 1,* $H/L = 2.00$	*Case 3,* $H/L = 1.922$	*Case 2,* $H/L = 1.847$
Oleic	193	187	181
Linoleic	105	79	61
Abietic	329	390	470
Total	627	656	712

would be lower for case 1. The optimum value of H/L would depend upon the relative costs of stages and of solute–solvent separation.

3-13. Unsymmetrical design for a symmetrical split

When $\mathscr{T}_A/\mathscr{B}_A = \mathscr{B}_B/\mathscr{T}_B$, the split is symmetrical and cases 1, 2, and 3 become identical, with center feed and with H/L as given by Eq. (3-64). It follows that this symmetrical design will correspond to the minimum number of stages. If components A and B are the only solute components present and if they are present in equal amounts in the feed stream, the symmetrical design will also correspond to the minimum solute buildup at the feed stage, and thus approximately to the minimum solvent requirements. On the other hand, if the feed stream contains substantially different amounts of the two key components or if it contains other components, then the minimum solvent requirements will generally correspond to an unsymmetrical design, with the feed off-center and with H/L different from that given by Eq. (3-64), even though the $\mathscr{T}/\mathscr{B}$ values for the two key components are reciprocals of each other. In such cases, the optimum design may be an unsymmetrical cascade, if solvent-recovery costs are large relative to the cost of the extractor stages.

3-14. Multiple feeds

When several solute feed streams are introduced into the cascade, Eqs. (3-51) and (3-52) are employed for each solute component. As written, these equations describe a cascade with two solute feed streams, but their generalization to include more feed streams is obvious. Generally speaking, a great many solutions, corresponding to different values of N, G, F, and H/L, can be found which will all achieve a given separation. Only one of these represents the solution corresponding to the minimum total number of stages in the cascade, and that design will not necessarily correspond to the minimum cost. The use of these equations will be illustrated in the following example problem.

EXAMPLE PROBLEM 3-12 Two solute feed streams containing linoleic and abietic acids are to be separated in a fractional-extraction column employing heptane as the light solvent and methyl Cellosolve containing 10 per cent water by volume as the heavy solvent. The solvent streams will be solute free when they are fed to the cascade. Each of the two solute feed streams is to be fed at a rate of 100 lb/hr. One of these feed streams contains 80 mass per cent linoleic and 20 mass per cent abietic acid; the other solute feed stream contains 20 mass per cent linoleic acid and 80 mass per cent abietic acid. It is desired to produce a top product containing linoleic and abietic acids in the mass ratio of 98:2 and a bottom product containing linoleic and abietic acids in the mass ratio of 2:98. What cascade design will minimize the total number of stages?

Solution: The material-balance table, Table 3-7, shows the mass flow rates of the two acids in the various feed and product streams. Since the solvent feed streams are solute free, the last two terms in Eqs. (3-51) and (3-52) are absent. Both of these equations must be satisfied for each component, but it should be remembered that

Table 3-7

	Mass Flow Rates (lb/hr) in			
Component	*Feed "F"*	*Feed "G"*	*Top Product*	*Bottom Product*
Linoleic acid	20	80	98	2
Abietic acid	80	20	2	98
Total	100	100	100	100

Eqs. (3-51) and (3-52) and the overall material balance, Eq. (3-50), are not all independent; any two of these three equations imply the third. Since the mass flow rates appearing in Table 3-7 have been chosen to satisfy Eq. (3-50), it is not necessary to use both Eqs. (3-51) and (3-52). Either of these equations could be used in principle, but the precision of the calculation would be poor when applying Eq. (3-51) to linoleic acid, and the precision of the calculation would be poor when applying Eq. (3-52) to abietic acid. Thus it is preferable to apply the equation for the top product when dealing with abietic acid and to apply the equation for the bottom product when dealing with linoleic acid. Generally, one should use the equation applying to the product stream that contains very little of the component in question rather than using the equation that applies to the product stream that contains most of that component.

Applying Eq. (3-51) for abietic acid yields

$$2\left(\frac{1}{E_{\mathrm{B}}} - E_{\mathrm{B}}^{N}\right) = 80\left(E_{\mathrm{B}}^{N-F} - E_{\mathrm{B}}^{N}\right) + 20\left(E_{\mathrm{B}}^{N-G} - E_{\mathrm{B}}^{N}\right)$$

For linoleic acid, Eq. (3-52) becomes

$$2\left(E_{\mathrm{A}} - \frac{1}{E_{\mathrm{A}}^{N}}\right) = 20\left(\frac{1}{E_{\mathrm{A}}^{F-1}} - \frac{1}{E_{\mathrm{A}}^{N}}\right) + 80\left(\frac{1}{E_{\mathrm{A}}^{G-1}} - \frac{1}{E_{\mathrm{A}}^{N}}\right)$$

It is apparent from the symmetry of the feed and product streams that a symmetrical cascade design exists for which

$$\frac{H}{L} = \sqrt{(2.17)(1.57)} = 1.847$$

$$E_{\mathrm{A}} = \frac{1}{E_{\mathrm{B}}} = \sqrt{\frac{2.17}{1.57}} = 1.176$$

$$F - 1 = N - G$$

With this symmetrical design, Eq. (3-52) applied to linoleic acid and Eq. (3-51) applied to abietic acid become identical, and only one of these is required. This equation takes the form

$$2\left(1.176 - \frac{1}{1.176^{N}}\right) = 20\left(\frac{1}{1.176^{F-1}} - \frac{1}{1.176^{N}}\right) + 80\left(\frac{1}{1.176^{G-1}} - \frac{1}{1.176^{N}}\right)$$

For a first approximation, the terms 1.176^{-N} can be neglected, giving

$$2.352 = \frac{20}{1.176^{F-1}} + \frac{80}{1.176^{G-1}}$$

There are many solutions to this equation. Arbitrarily choosing F equal to 21, this equation can be solved for $G = 25.25$. The total number of stages is then calculated as

$$N = F + G - 1 = 45.25$$

Now it is possible to check to see if 1.176^{-N} can be neglected or if a small correction is needed. Inserting $N = 45.25$ and $F = 21$ into the complete equation results in the solution $G = 25$, and thus $N = 45$. This solution was based upon the arbitrary choice $F = 21$. With other choices for F other solutions can be obtained in the same manner, giving the results shown in Table 3-8.

Table 3-8

F	G	N	*Remarks*
14.2	∞	∞	Minimum possible F
15	35.1	49.1	
16	30.7	45.7	
17	28.5	44.5	
18	27.2	44.2	Minimum total stages
19	26.2	44.2	Minimum total stages
20	25.6	44.6	
21	25	45	
22	24.6	45.6	
23	24.3	46.3	
24	24	47	Both feeds to middle stage; Example Problem 3-7
28	23.4	50.4	
∞	22.8	∞	Minimum possible G

The minimum total number of stages for the cascade corresponds to a solution between $F = 18$ and $F = 19$. This can be checked with the aid of differential calculus. For a symmetrical cascade of this type, Eq. (3-51) can be written

$$\mathscr{T}\left(\frac{1}{E} - E^N\right) = \mathscr{F}(E^{N-F} - E^N) + \mathscr{G}(E^{F-1} - E^N)$$

because $N - G = F - 1$. Assume that the value of N is fixed and seek the value of F that will maximize $\mathscr{T}$. Differentiating gives

$$\left(\frac{1}{E} - E^N\right)\frac{d\mathscr{T}}{dF} = -\mathscr{F}E^{N-F}\ln E + \mathscr{G}E^{F-1}\ln E$$

This derivative will be zero when

$$\frac{\mathscr{G}}{\mathscr{F}} = E^{N+1-2F} = E^{G-F}$$

In this case, for linoleic acid, $\mathscr{G}/\mathscr{F} = \frac{80}{20} = 4$, and $E = 1.176$. Therefore,

$$N + 1 - 2F = G - F = 8.6$$

Taking N to be 44.2, this gives $F = 18.3$, $G = 26.9$, which is surely consistent with the values given in Table 3-8.

The solution with $F = 24$, $G = 24$, corresponds to mixing the two feed streams together and feeding them both into the middle stage of the cascade, which is the design of Example Problem 3-7. On the other hand, for the present case, in which the two solute feed streams already represent a certain degree of separation between linoleic and abietic acids, the cascade that corresponds to mixing these two feed streams and sending them to the same stage will require a greater number of stages than will the cascade design that corresponds to $F = 19$, $G = 26.2$. Note that for this design feed stream "F," the feed rich in abietic acid, is fed into the cascade beneath the feed point for feed stream "G." The last two entries in the table correspond to feeding feed stream "G" beneath the feed point for feed stream "F." Thus the feed stream rich in linoleic acid is fed near the bottom of the cascade, where the product stream is to be rich in abietic acid. This makes the separation more difficult, and the number of stages required for the cascade exceeds the number required if the two feed streams were mixed together and sent to the middle stage.

The results of Table 3-8 correspond to symmetrical designs for this cascade in which the extraction factors for the two components are reciprocals of each other and in which the feed and product streams are arranged symmetrically. It would be possible to design unsymmetrical cascades to accomplish this separation. The total number of stages required would exceed 44.2, but the total cost might be lower in some cases.

FRACTIONAL EXTRACTION WITH REFLUX

When the light- and heavy-solvent feed streams contain the solutes, the last two terms in Eqs. (3-51) and (3-52) will not be zero and the complete equations must be used. An example of this type of operation is a fractional-extraction cascade with solute reflux. A diagram of such a system with a single solute feed stream is shown in Fig. 3-13. The light-solvent stream leaving the top stage in the cascade is sent to a light-solvent separator where the light solvent and the accompanying solutes are separated from each other, typically by vaporizing the solvent. The light solvent is then returned to the bottom of the cascade, and the solute stream is divided into a top product stream and a top reflux stream. This reflux stream is dissolved in the heavy solvent before it is sent to the top stage in the cascade. Similarly, the heavy-solvent stream leaving the bottom stage of the cascade is sent to a heavy-solvent separator, where the heavy solvent is separated from the solutes. The heavy solvent is returned to the top of the cascade, and the bottom solute stream is divided into a bottom product stream and a bottom reflux stream. The bottom reflux stream is mixed with the light-solvent stream before it is sent to the bottom stage of the cascade.

The use of reflux corresponds to returning a portion of the top and bottom product streams to the stage in the cascade from which they came, but dissolved in the other solvent stream. The effect of refluxing a portion of the product streams is to decrease the number of stages required to effect a given separation. On the other hand, the solute flow rates throughout the cascade are increased by the use of reflux, and this will generally require larger flow rates of the solvent streams to keep the solute concentrations in the solvent streams acceptably low. Higher solvent flow rates will generally require larger contacting equipment for each stage and will generally result in higher solvent-separation costs. These compensating effects will usually dictate an optimum amount of reflux which will minimize the total cost of the extraction operation. For a more detailed consideration of the use of

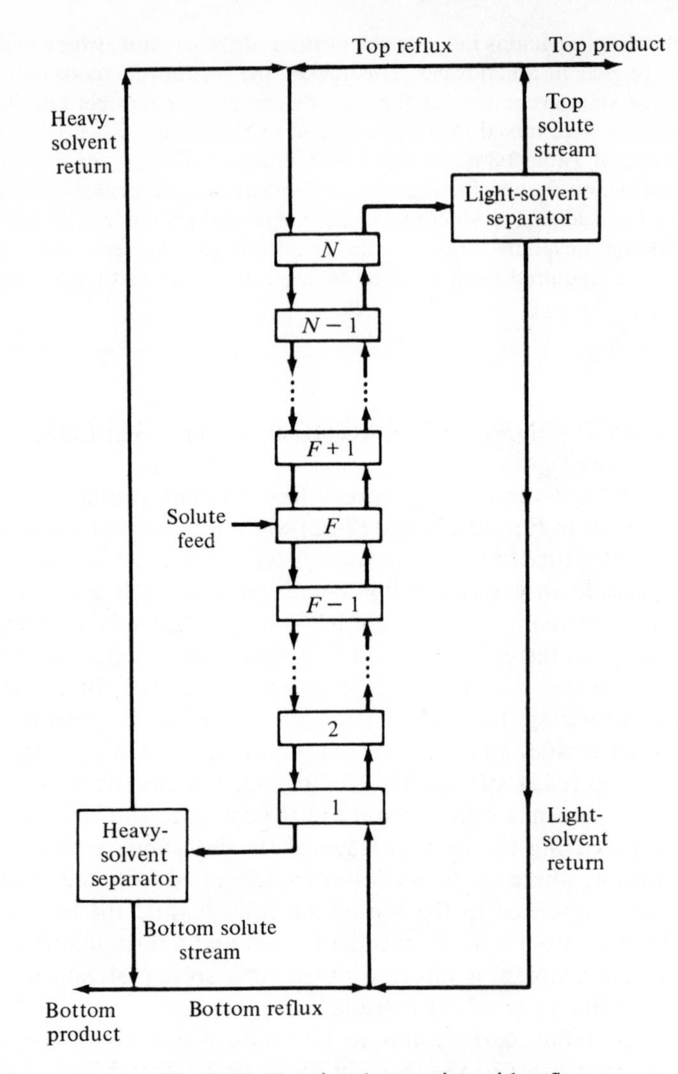

Fig. 3-13. Fractional extraction with reflux.

reflux, the reader may consult references [2, 11]. A brief treatment will be given here.

Consider the system shown in Fig. 3-13. For simplicity it will be assumed that the light-solvent separator achieves a perfect separation between the light solvent and the solutes; thus the top solute stream will contain all the solutes carried by the light-solvent stream from the top stage of the cascade. The fraction of the top-solute stream returned as top reflux will be denoted by ϵ_T; the remaining fraction will be withdrawn as the top-product stream.

The *reflux ratio* is defined as the ratio of the reflux rate to the product rate; the top reflux ratio is given by

$$\mathscr{R}_T = \frac{\epsilon_T}{1 - \epsilon_T} \tag{3-75}$$

It will also be assumed that the heavy-solvent separator achieves a perfect separation between the heavy solvent and the solutes. Similarly, the fraction of the bottom solute stream returned as the bottom reflux stream will be denoted ϵ_B with the remainder of that stream being withdrawn as the bottom product. Thus the bottom reflux ratio is given by

$$\mathscr{R}_B = \frac{\epsilon_B}{1 - \epsilon_B} \tag{3-76}$$

With these assumptions and definitions, it follows that, for each solute component, the rate at which the solute enters the cascade with the heavy-solvent stream fed to the top stage is simply the fraction ϵ_T of the rate at which that solute leaves the top stage of the cascade with the light-solvent stream:

$$\mathscr{H} = \epsilon_T \mathscr{T} \tag{3-77}$$

Similarly, the rate of flow of a solute component into the bottom stage of the cascade with the light-solvent stream is the fraction ϵ_B of the rate at which that solute component leaves the bottom stage of the cascade with the heavy solvent:

$$\mathscr{L} = \epsilon_B \mathscr{B} \tag{3-78}$$

Inserting Eqs. (3-77) and (3-78) into Eq. (3-50), the overall material balance for any solute component may be written as

$$(1 - \epsilon_B)\mathscr{B} + (1 - \epsilon_T)\mathscr{T} = \mathscr{F} \tag{3-79}$$

Substitution of Eqs. (3-77), (3-78), and (3-79) into Eq. (3-51) yields the result

$$\frac{\mathscr{T}}{\mathscr{B}} = \frac{(1 - \epsilon_B)E^{N-F} + \epsilon_B E^{N-1} - E^N}{1/E - (1 - \epsilon_T)E^{N-F} - \epsilon_T} \tag{3-80}$$

Equation (3-80) describes the effect of solute reflux upon the number of stages required to achieve a given separation in a fractional-extraction cascade. If no solute reflux is employed at either end of the cascade, ϵ_T and ϵ_B are both zero, and Eq. (3-80) reduces to Eq. (3-55). At the other extreme, as the reflux ratios at the top and the bottom of the cascade approach infinity, ϵ_T and ϵ_B approach unity, and Eq. (3-80) approaches the simple form

$$\frac{\mathscr{T}}{\mathscr{B}} = E^N \qquad \text{(for total reflux)} \tag{3-81}$$

With a fixed solute-feed-stream flow rate, reflux ratios approaching infinity imply solute flow rates throughout the cascade approaching infinity; therefore, this is a case never achieved in practice. But the same effect can be achieved in a real cascade by shutting off the solute-feed-stream flow and shutting off the top and bottom product withdrawal flows, thus simply returning all solute products as reflux. The concentration profile in a cascade run in this fashion would correspond to the idealized limit of infinite reflux rates with finite solute feed and product rates; therefore, this limiting case is often referred to as *total reflux*. Remembering Eq. (3-56) for a cascade fed at the middle stage, it is clear from Eq. (3-81) that the number of stages in the cascade can be reduced by approximately a factor of 2 by employing top and bottom reflux ratios approaching infinity.

EXAMPLE PROBLEM 3-13 Referring to Example Problem 3-7, how many stages would be required if the cascade design were modified to include solute reflux with top and bottom reflux ratios both equal to 4:1? Repeat the calculation for ratios of 9:1 and 19:1.

Solution: With reflux ratios of 4:1 at the top and at the bottom, ϵ_T and ϵ_B are equal to 0.8. Because of the symmetry of the feed-stream composition and the split, a symmetrical design will correspond to the minimum number of total stages. Therefore, as in Example Problem 3-7, $H/L = 1.847$, the solute feed stream will be introduced into the middle stage, and

$$E_{\text{A}} = \frac{1}{E_{\text{B}}} = 1.176$$

Since $\epsilon_T = \epsilon_B$, the value of $\mathscr{T}/\mathscr{B}$ for each component is equal to the ratio of its flow rate in the top product stream to its flow rate in the bottom product stream. For species A, linoleic acid, Eq. (3-80) becomes

$$\frac{98}{2} = \frac{(1.176)^N - 0.8(1.176)^{N-1} - 0.2(1.176)^{N-F}}{0.2(1.176)^{N-F} + 0.8 - 1/1.176}$$

Since the solute feed will be introduced into the middle stage,

$$N - F = F - 1$$

By trial and error the solution is found to be $F = 21.2$, $N = 41.4$. Similar calculations for other reflux ratios yield the results shown in Table 3-9. The first entry in this table is the result of Example Problem 3-7 in which no solute reflux was used. The last entry in the table follows from Eq. (3-81).

Table 3-9

$\mathscr{R}_T = \mathscr{R}_B$	$\epsilon_T = \epsilon_B$	F	N
0	0	24	47
4/1	0.8	21.2	41.4
9/1	0.9	19	37
19/1	0.95	16.7	32.4
∞	1	12.5	24

3-15. Effect of reflux upon solute buildup

The use of solute reflux at the top and the bottom of the cascade results in increased solute flow rates throughout the cascade. The percentage increase is greatest at the top and at the bottom of the cascade; the highest total solute flow rates may still occur at the feed stage, or at high reflux ratios they may occur at the top or bottom stage. The extent of the solute buildup can be computed with Eq. (3-67). Multiplying Eq. (3-67) through by H and employing Eqs. (2-56), (3-46), and (3-47) yields

$$HX_F = \frac{1 - E^F}{1 - E}\mathscr{B} - \frac{1 - E^{F-1}}{1 - E}\mathscr{L} \tag{3-82}$$

The quantity on the left-hand side of Eq. (3-82) represents the molal flow rate of the solute component in the heavy-solvent stream leaving the feed stage. Employing Eq. (3-78), which assumes that the light- and heavy-solvent separators effect perfect separations, Eq. (3-82) becomes

$$\frac{HX_F}{\mathscr{B}} = \frac{1 - E^F}{1 - E} - \epsilon_B \frac{1 - E^{F-1}}{1 - E} \tag{3-83}$$

The solute flow rate into the feed stage with the light-solvent stream flowing from stage $F - 1$ is computed from a material balance around the bottom $F - 1$ stages of the cascade:

$$HX_F + \mathscr{L} = LY_{F-1} + \mathscr{B} \tag{3-84}$$

Combining Eqs. (3-78) and (3-84) yields

$$\frac{LY_{F-1}}{\mathscr{B}} = \frac{HX_F}{\mathscr{B}} + \epsilon_B - 1 \tag{3-85}$$

Equations (3-83) and (3-85) apply also to any stage beneath the feed stage if F is replaced by j, the stage number.

The equations for the top of the cascade corresponding to Eqs. (3-83) and (3-85) can be inferred by symmetry:

$$\frac{LY_F}{\mathscr{T}} = \frac{1 - 1/E^{N-F+1}}{1 - 1/E} - \epsilon_T \frac{1 - 1/E^{N-F}}{1 - 1/E} \tag{3-86}$$

$$\frac{HX_{F+1}}{\mathscr{T}} = \frac{LY_F}{\mathscr{T}} + \epsilon_T - 1 \tag{3-87}$$

Equations (3-86) and (3-87) apply also to any stage above the feed stage if F is replaced by j, the stage number.

The use of these equations and their implications are best illustrated by means of the following example problem.

EXAMPLE PROBLEM 3-14 For the cascade design of Example Problem 3-13 with top and bottom reflux ratios of 9:1, calculate the solute flow rates entering and leaving the feed stage for the case of a solute feed rate to the cascade of 1000 lb/hr of the mixture of linoleic and abietic acids. What must be the heavy- and light-solvent flow rates if the maximum acid concentration of either solvent phase at any point within the cascade should not exceed 0.08 lb of total acid per pound of solvent?

Solution: The overall material balances for the case of top and bottom reflux ratios of 9:1 are easily solved for the flow rates of the external streams. The mass flow rates are summarized in Table 3-10. Note that the mass flow rates of the solutes

Table 3-10

	Mass Flow Rate (lb/hr) of		
Stream	*Linoleic Acid*	*Abietic Acid*	*Total Acids*
Solute feed	500	500	1000
Top product	490	10	500
Top reflux = $\mathscr{H}$	4410	90	4500
$\mathscr{T}$	4900	100	5000
Bottom product	10	490	500
Bottom reflux = $\mathscr{L}$	90	4410	4500
$\mathscr{B}$	100	4900	5000

leaving the top stage of the cascade with the light-solvent stream are 10 times as great as they were in Example Problems 3-7 and 3-10 because of the 9:1 reflux ratio in this case. The same is true of the solute flow rates leaving the bottom stage of the cascade with the heavy-solvent stream.

The mass flow rate of linoleic acid in the heavy-solvent stream leaving the feed stage, stage 19, is given by Eq. (3-83).

$$\frac{HX_{F_A}}{100} = \frac{1.176^{19} - 1}{0.176} - 0.9\frac{1.176^{18} - 1}{0.176}$$

The result is

$$HX_{F_A} = 2840$$

For abietic acid, Eq. (3-83) becomes

$$\frac{HX_{F_B}}{4900} = \frac{1 - 1/1.176^{19}}{1 - 1/1.176} - 0.9\frac{1 - 1/1.176^{18}}{1 - 1/1.176}$$

with the result

$$HX_{F_B} = 3360$$

The solute contents of the light-solvent stream entering the feed stage from stage $F - 1$ are given by Eq. (3-85). For linoleic acid and abietic acid, respectively, Eq. (3-85) gives

$$LY_{F-1_A} = 2840 - (0.1)(100) = 2830$$

$$LY_{F-1_B} = 3360 - (0.1)(4900) = 2870$$

For the light-solvent stream leaving the feed stage, Eq. (3-86) is employed.

For linoleic acid

$$\frac{LY_{F_A}}{4900} = \frac{1 - 1/1.176^{19}}{1 - 1/1.176} - 0.9\frac{1 - 1/1.176^{18}}{1 - 1/1.176}$$

with the result

$$LY_{F_A} = 3360$$

For abietic acid

$$\frac{LY_{F_B}}{100} = \frac{1.176^{19} - 1}{0.176} - 0.9\frac{1.176^{18} - 1}{0.176}$$

with the result

$$LY_{F_B} = 2840$$

The solute contents of the heavy-solvent stream entering the feed stage from above are given by Eq. (3-87). For linoleic acid

$$HX_{F+1\,A} = 3360 - (0.1)(4900) = 2870$$

For abietic acid

$$HX_{F+1\,B} = 2840 - (0.1)(100) = 2830$$

These solute flow rates around the feed stage are summarized in Table 3-11. From this table and Table 3-10 the symmetry of the system is apparent.

Table 3-11

	Mass Flow Rate (lb/hr) of		
Stream	*Linoleic Acid*	*Abietic Acid*	*Total Acids*
HX_F	2840	3360	6200
LY_F	3360	2840	6200
HX_{F+1}	2870	2830	5700
LY_{F-1}	2830	2870	5700

Equation (3-83) can be generalized to apply to any stage beneath the feed stage. For the second stage,

$$\frac{HX_2}{\mathscr{B}} = \frac{1 - E^2}{1 - E} - \epsilon_B$$

and for stage 18,

$$\frac{HX_{18}}{\mathscr{B}} = \frac{1 - E^{18}}{1 - E} - \epsilon_B\frac{1 - E^{17}}{1 - E}$$

Similar equations can be written for other stages. The calculations are summarized in Table 3-12.

It is seen that the linoleic acid flow rate in the heavy-solvent stream decreases from 2840 at the feed stage to 100 at the bottom stage, while the abietic acid flow rate increases from 3360 at the feed stage to 4900 at the bottom stage. The total acid flow rate goes through a minimum. The highest total acid flow rate occurs at the feed

Table 3-12

	Mass Flow Rate (lb/hr) of		
Stream	*Linoleic Acid*	*Abietic Acid*	*Total Acids*
$HX_F = HX_{19}$	2840	3360	6200
HX_{18}	2420	3380	5800
HX_{10}	620	3680	4300
HX_2	128	4630	4758
$HX_1 \equiv \mathscr{B}$	100	4900	5000

stage, but at higher reflux ratios it would occur at the bottom stage. In this symmetrical case, a similar pattern exists in the light-solvent stream above the feed, with the roles of the two acids interchanged.

At this reflux ratio of 9:1, the highest total acid flow rate occurs in the two streams leaving the feed stage. Since $H/L = 1.847$, the total acid concentration is greatest in the light solvent leaving the feed stage:

$$Y_{F_A} + Y_{F_B} = \frac{6200}{L}$$

To keep $Y_{F_A} + Y_{F_B}$ less than 0.08, the light-solvent flow rate must be at least equal to

$$L = \frac{6200}{0.08} = 77{,}500 \text{ lb/hr}$$

The coresponding heavy-solvent flow rate is

$$H = (77{,}500)(1.847) = 143{,}200 \text{ lb/hr}$$

These solvent flow rates are 5 per cent higher than those required without reflux, as determined in Example Problem 3-10.

Referring to Example Problem 3-13, it was seen that top and bottom reflux ratios of 9:1 resulted in a 21 per cent reduction in the total number of stages required in the cascade, as compared to the case of no reflux. Comparing the results obtained here with those of Example Problem 3-10, it is seen that the use of this reflux increased the required light- and heavy-solvent flow rates by 5 per cent, because the total solute flow rates in the middle of the cascade increased by that amount. The solute flow rates at other points within the cascade increased by a much larger percentage, but the maximum total solute flow rate still occurs at the feed stage. It is seen, therefore, in this example that the use of this amount of reflux results in a 21 per cent reduction in N while increasing the solvent flow rates only 5 per cent. This probably represents an economic saving, although the relative costs of contacting equipment versus solvent recovery would have to be known to assess this point.

However, it should be pointed out that the criterion adopted in this example problem, that of keeping the *total* acid content of the solvent streams less than a specified value, might not be a good criterion in all cases. In this regard it should be noted that the linoleic acid flow rate in the light-solvent stream leaving the top stage is 4900 lb/hr, substantially higher than the 3360 lb/hr leaving the feed stage with the light-solvent stream. Likewise, the abietic acid flow rate in the heavy-solvent stream leaving the bottom stage is 4900 lb/hr versus only 3360 lb/hr in the heavy-solvent stream leaving the feed stage. If it should happen that one of the solute components is much more effective than the other in promoting the mutual solubility of the solvent components, it might be that refluxing that component would result in much greater solvent flow rate increases than those indicated in this simple example problem.

3-16. Effect of reflux on cascade performance

The preceding discussion and example problems by no means represent an exhaustive study of the effects of reflux, but they do reveal the essential elements. The use of solute reflux will result in a decrease in the number of stages required but will increase solvent-circulation rates. This later effect is easily understood, but the effect of reflux in decreasing the required number of stages is not readily visualized, although it surely follows from Eq. (3-80).

The McCabe–Thiele diagram is a very useful graphical aid to visualizing the effect of reflux upon the number of stages required in a binary distillation column, and this subject is discussed in Chap. 4. After studying Chap. 4, the reader may wish to return to a consideration of the effect of reflux in the fractional extraction of a binary solute stream using two solvents, as in Example Problems 3-13 and 3-14. This will be greatly aided by constructing a McCabe–Thiele diagram on a solvent-free basis. Thus $Y_A/(Y_A + Y_B)$ is plotted versus $X_A/(X_A + X_B)$. The equilibrium curve takes the same form as it does in distillation, and the chord slopes of the curved operating lines are given by the ratios of the total solute flow rates in the rising and descending streams. These total solute flow rates are obviously increased by the use of reflux, especially the flow rates near the top and bottom of the cascade, and this results in moving the operating lines away from the equilibrium curve. The result is a decrease in the number of stages required, just as in distillation.

EFFECT OF SOLVENT ADDITION TO THE FEED STAGE

Throughout this discussion of fractional-extraction cascades it has been assumed that no solvent accompanied the solute feed to the feed stage. Thus all the light solvent has been assumed to enter at the bottom stage and all the heavy solvent at the top stage. This results in L/H being constant throughout the cascade and simplifies the development of a general description, such as Eqs. (3-51) and (3-52). This simplifying assumption will, however, not always apply.

Consider a fractional-extraction cascade with solute-free heavy and light solvents fed to the top and bottom stages, respectively, and with one solute feed stream entering at stage F. If either or both solvents also enter at the feed stage in appreciable quantities, the solvent-flow-rate ratio L/H will be different in the upper and lower sections of the cascade, above and below the feed stage, and so will the value of E for any component. Three extraction factors will be defined for any given component: E_U will employ the L/H value in the upper section of the cascade; E_L will employ the L/H value in the lower section of the cascade; E_F will employ the L and H values *leaving* the feed stage, that is, L for the upper section and H for the lower section. In terms of these definitions, the cascade performance is described by

$$\frac{\mathscr{T}}{\mathscr{B}} = \frac{E_F[(E_L)^F - 1](1 - 1/E_U)}{(E_L - 1)[1 - (1/E_U)^{N-F+1}]} \tag{3-88}$$

which reduces to Eq. (3-55) when L/H is constant throughout the cascade and $E_U = E_L = E_F = E$. Equation (3-88) is most readily derived by applying Eq. (3-68) to the section of the cascade below the feed stage, applying Eq. (3-69) to the section above the feed stage, and combining the results with the equilibrium expression for the feed stage. The derivation is left as an exercise for the reader.

The result of adding either or both solvents to the feed stage is to degrade the separation achieved with a given number of stages or to increase the number of stages required for a given separation. This is readily understood, because solvent addition to the feed stage decreases E_L and increases E_U. Thus the lower section of the cascade cannot as readily remove species A from the bottom product because E_L is decreased, and the upper section of the cascade cannot as readily remove species B from the top product because $1/E_U$ is also decreased.

If the percentage of the total solvent flow rates added at the feed stage is too large, relative to the percentage difference between D_A and D_B, a sharp separation between components A and B cannot be achieved, even with an infinite number of stages in the cascade. On the other hand, solvent addition to the feed stage is not always detrimental. Addition of a moderate percentage of the total solvent flow rates to the feed stage will increase the number of stages required for a given separation but will also decrease the solute buildup at the feed stage. If solvent separation costs are large, the net result may be economically advantageous.

It is sometimes asserted that operating with a solvent-free solute feed stream is difficult or impossible. For example, when the solute components are solid materials, it is natural to wish to dissolve them in one of the solvents and feed the solution to the cascade. But this extra fresh solvent addition at the feed stage can be avoided by withdrawing the heavy-phase underflow from stage $F + 1$, dissolving the solute feed material in this liquid, and returning this stream to stage F. Alternatively, the light-phase overflow from stage $F - 1$ can be used to dissolve the solute feed material, and this liquid can be returned to the feed stage. With this scheme no additional light or heavy solvent need be introduced at the feed stage, and the decision to add solvent at the feed stage can be based on economic considerations.

Since solvent addition at the feed stage increases the number of stages required to achieve a specified separation, it is to be expected that solvent removal from the feed stage would have the opposite effect, as indeed it does. Some of the light solvent can be removed from the light phase rising from stage $F - 1$ to the feed stage, and some of the heavy solvent can be removed from the heavy phase descending from stage $F + 1$ to the feed stage. This can be accomplished by the same solvent–solute separation methods as those used with the top and bottom product streams. Equation (3-88) describes this system also, but solvent removal from the feed stage results in an increase in E_L and a decrease in E_U, allowing a given separation to be accomplished with fewer stages. On the other hand, the solute buildup at the feed stage will be increased, requiring greater total solvent-circulation rates, and the net effect may not be economically advantageous.

These interactions are illustrated in the following example problems.

EXAMPLE PROBLEM 3-15 Repeat Example Problems 3-7 and 3-10 with the following changes:

(a) Ten per cent of each solvent will enter at the feed stage.
(b) Twenty per cent of each solvent will enter at the feed stage.
(c) Ten per cent of each solvent will be removed at the feed stage.
(d) Twenty per cent of each solvent will be removed at the feed stage.

Solution: Symmetry will be preserved with solute feed at the center and with $H/L = 1.847$, where H and L refer to the *total* flow rate of each solvent to the cascade.

(a) The extraction factors for the various sections of the cascade are

Acid	E_U	E_F	E_L
Linoleic	$1.176/0.9 = 1.3065$	1.176	$(0.9)(1.176) = 1.0584$
Abietic	$1/(0.9)(1.176) = 0.9451$	$1/1.176 = 0.8507$	$0.9/1.176 = 0.7656$

With center feed, $N - F + 1 = F$, and Eq. (3-88) becomes, for linoleic acid,

$$49 = \frac{(1.176)[(1.0584)^F - 1](1 - 1/1.3065)}{(1.0584 - 1)[1 - (1/1.3065)^F]}$$

and for abietic acid

$$\frac{1}{49} = \frac{(0.8507)[1 - (0.7656)^F](1 - 1/0.9451)}{(1 - 0.7656)[1 - (1/0.9451)^F]}$$

These two equations are identical, and a trial-and-error solution of either of them yields $F = 42.9$, or $N = 84.8$.

The solute buildup at the feed stage is obtained by using Eq. (3-69) with E_U. For linoleic acid

$$(LY_F)_A = \frac{(490)[1 - (1/1.3065)^{42.9}]}{1 - (1/1.3065)} = 2090 \text{ lb/hr}$$

For abietic acid

$$(LY_F)_B = \frac{(10)[(1/0.9451)^{42.9} - 1]}{(1/0.9451) - 1} = 1780 \text{ lb/hr}$$

Summing these values gives

$$(LY_F)_A + (LY_F)_B = 3870 \text{ lb/hr}$$

and, therefore,

$$L = \frac{3870}{0.08} = 48{,}600 \text{ lb/hr}$$

assuming that the light-solvent stream leaving the feed stage will be the richest in total acids. Since the light-solvent stream entering the feed stage contains 10 per cent less light solvent, it should be checked also. The equilibrium relationship for the feed stage for any component can be written

$$HX_F = \frac{LY_F}{E_F}$$

For abietic acid, then

$$(HX_F) = \frac{1780}{0.8507} = 2090 \text{ lb/hr}$$

and for linoleic acid

$$(HX_F)_A = \frac{2090}{1.176} = 1780 \text{ lb/hr}$$

which emphasizes the symmetry of the cascade. Since the light solvent enters the bottom stage free of solutes, a material balance around the bottom $F - 1$ stages for abietic acid shows 2090 lb/hr entering stage $F - 1$ with the heavy solvent from the feed stage, 490 lb/hr leaving with the bottom product, and therefore 2090 − 490 = 1600 lb/hr leaving stage $F - 1$ with the light-solvent stream rising to the feed stage. A similar balance for linoleic acid shows 1770 lb/hr in the light-solvent stream leaving stage $F - 1$. The total acid content of the light solvent leaving stage $F - 1$ is 1600 + 1770 = 3370 lb/hr, which is 13 per cent lower than the total acid content of the light-solvent stream leaving the feed stage. This latter stream will therefore be the richest in total acids, and the required value of L is 48,600 lb/hr, as computed earlier. The required total heavy-solvent flow rate is

$$H = (1.847)(48{,}600) = 89{,}800 \text{ lb/hr}$$

These solvent flow rates are 34.5 per cent lower than the values computed in the solution to Example Problem 3-10. On the other hand, the 84.8 stages required here are 80 per cent greater than the result computed in the solution to Example Problem 3-7.

(b) The extraction factors are

Acid	E_U	E_F	E_L
Linoleic	1.176/0.8 = 1.470	1.176	(0.8)(1.176) = 0.9408
Abietic	1/(0.8)(1.176) = 1.0634	1/1.176 = 0.8507	0.8/1.176 = 0.6806

In this case, the percentage of each solvent added at the feed stage is greater than the percentage by which $\sqrt{\alpha}$ differs from unity, and this results in E_L being less than unity for linoleic acid and in E_U being greater than unity for abietic acid. This will prevent the attainment of the desired separation, as is seen below.

Equation (3-88) becomes, for linoleic acid,

$$49 = \frac{(1.176)[1 - (0.9408)^F][1 - (1/1.470)]}{(1 - 0.9408)[1 - (1/1.470)^F]}$$

and the corresponding equation for abietic acid is equivalent. The right-hand side of this equation approaches a maximum value of 6.35 as F approaches infinity, and it is seen that the specified separation, $\mathscr{T}_A/\mathscr{B}_A = \mathscr{B}_B/\mathscr{T}_B = 49$, cannot be attained.

(c) The extraction factors are

Acid	E_U	E_F	E_L
Linoleic	(0.9)(1.176) = 1.0584	1.176	1.176/0.9 = 1.3065
Abietic	0.9/1.176 = 0.7656	1/1.176 = 0.8507	1/(0.9)(1.176) = 0.9451

Equation (3-88) for linoleic acid becomes

$$49 = \frac{(1.176)[(1.3065)^F - 1](1 - 1/1.0584)}{(1.3065 - 1)[1 - (1/1.0584)^F]}$$

and the corresponding equation for abietic acid is equivalent. Solving by trial and error gives $F = 18.83$, or $N = 36.66$.

The solute buildup at the feed stage is obtained by using Eq. (3-69) with E_U. For linoleic acid

$$0.9LY_{F_A} = \frac{(490)[1 - (1/1.0584)^{18.83}]}{1 - (1/1.0584)} = 5840 \text{ lb/hr}$$

For abietic acid

$$0.9LY_{F_B} = \frac{(10)[(1/0.7656)^{18.83} - 1]}{(1/0.7656) - 1} = 4970 \text{ lb/hr}$$

Summing these gives

$$0.9L(Y_{F_A} + Y_{F_B}) = 10{,}810 \text{ lb/hr}$$

and therefore the light-solvent flow rate leaving the feed stage must be

$$0.9L = \frac{10{,}810}{0.08} = 135{,}000 \text{ lb/hr}$$

The total light-solvent rate is

$$L = \frac{135{,}000}{0.9} = 150{,}000 \text{ lb/hr}$$

and

$$H = (150{,}000)(1.847) = 277{,}000 \text{ lb/hr}$$

Compared with the design of Example Problems 3-7 and 3-10, this design is seen to correspond to a 22 per cent reduction in the number of stages but a 102 per cent increase in the solvent flow rates.

(d) The extraction factors are

Acid	E_U	E_F	E_L
Linoleic	(0.8)(1.176) = 0.9408	1.176	1.176/0.8 = 1.470
Abietic	0.8/1.176 = 0.6806	1/1.176 = 0.8507	1/(0.8)(1.176) = 1.0634

Equation (3-88) for linoleic acid is

$$49 = \frac{(1.176)[(1.470)^F - 1][(1/0.9408) - 1]}{(1.470 - 1)[(1/0.9408)^F - 1]}$$

and the corresponding equation for abietic acid is equivalent. Solving gives $F = 16.3$, and $N = 31.6$.

Solute flow rates at the feed stage are

$$0.8LY_{F_A} = \frac{(490)[(1/0.9408)^{16.3} - 1]}{(1/0.9408) - 1} = 13{,}300 \text{ lb/hr}$$

$$0.8LY_{F_B} = \frac{(10)[(1/0.6806)^{16.3} - 1]}{1/0.6803 - 1} = 11{,}300 \text{ lb/hr}$$

Summing these gives 24,600 lb/hr of total acids, and

$$0.8L = \frac{24{,}600}{0.08} = 308{,}000 \text{ lb/hr}$$

$$L = 385{,}000 \text{ lb/hr}$$

$$H = (385{,}000)(1.847) = 711{,}000 \text{ lb/hr}$$

Compared with the design of part (c), this design represents a small decrease in N and a large increase in L and H.

EXAMPLE PROBLEM 3-16 Repeat Example Problems 3-7 and 3-10 with the following changes:

(a) Twenty per cent of the heavy solvent will enter at the feed stage, the rest entering at the top.

(b) Forty per cent of the heavy solvent will enter at the feed stage, the rest entering at the top.

In both cases, all the light solvent will enter at the bottom stage.

Solution: In this problem the symmetry is destroyed, and H/L values greater than 1.847 are expected to yield best results (H will refer to the *total* heavy solvent fed to the cascade).

(a) The extraction factors are

Acid	E_U	$E_F = E_L$
Linoleic	$(2.17/0.8)(L/H) = 2.7125(L/H)$	$2.17(L/H)$
Abietic	$(1.57/0.8)(L/H) = 1.9625(L/H)$	$1.57(L/H)$

For H/L values greater than 2.17, E_F and E_L for linoleic acid will be less than unity, and for H/L values less than 1.9625, E_U for abietic acid will be greater than unity; in either case, the desired sharp separation could not be achieved. For H/L values between 1.9625 and 2.17, corresponding to L/H values between 0.4605 and 0.51, various designs are possible. For a given value of L/H the extraction factors are readily computed, and these values are then substituted into Eq. (3-88), written for linoleic acid and for abietic acid, together with the specified values of $\mathscr{T}/\mathscr{B}$. The two resulting equations are solved simultaneously for F and $N - F + 1$ by a procedure similar to that illustrated in the solution to Example Problem 3-11. The solute buildup at the feed stage and the corresponding solvent requirements are then calculated as before. The results of such calculations are shown in Table 3-13.

Table 3-13

L/H	F	$N - F + 1$	N	L	H
0.47	85.886	31.366	116.252	45,638	97,102
0.48	55.072	38.913	92.985	45,139	94,039
0.4855	46.575	44.932	90.507	45,087	92,868
0.49	41.512	51.561	92.073	45,159	92,162
0.50	33.687	74.486	111.173	45,675	91,350

The design for $L/H = 0.4855$ corresponds to the minimum N and also to the minimum total solute flow rate rising from the feed stage, and therefore to minimum L. This design is not very different from that of part (a) of the solution to Example Problem 3-15. Compared to the design with no solvent added at the feed stage, the present design corresponds to a 93 percent increase in N, a 39 per cent decrease in L, and a 32 per cent decrease in H.

(b) The extraction factors are

Acid	E_U	$E_F = E_L$
Linoleic	$(2.17/0.6)(L/H) = 3.6167(L/H)$	$2.17(L/H)$
Abietic	$(1.57/0.6)(L/H) = 2.6167(L/H)$	$1.57(L/H)$

It is obvious that no value of L/H can simultaneously make E_L and E_F greater than unity for linoleic acid and make E_U less than unity for abietic acid. In this case, the specified separation cannot be achieved.

UTILITY OF THE LINEAR MODEL

This chapter has presented a linearized analysis of liquid–liquid extraction. The assumptions of constant distribution coefficients and constant solvent flow rates in the two phases are necessary for linear behavior, and these assumptions will generally not be valid at high solute concentrations. Furthermore, as discussed previously, the optimum solvent flow rates for separation by fractional extraction are dictated by the nonlinear behavior at high solute concentrations; thus a consideration of the nonlinear effects is important to the economic balance.

For the simple extraction of a single solute from one solvent phase to another, the nonlinear problem is usually solved by a graphical method on a triangular diagram; the reader is referred to references [15, 16] for a description of the method. But for a fractional-extraction separation involving several solute species, no such method is generally applicable. In such cases, the nonlinear equations describing the behavior of the various stages must be solved by iterative techniques broadly similar to those considered in Chap. 5 for multicomponent distillation systems. A digital computer will generally be used to accomplish the iterative solution. But a great deal of physical-chemical data are required for the nonlinear description of a multicomponent fractional-extraction system. The various distribution coefficients and also the mutual solubilities of the solvent components will generally depend upon the concentrations of all solute components, and an enormous amount of experimental data is usually required to describe these nonlinear effects. Therefore, it is often less expensive to run pilot-plant experiments on the multicomponent fractional-extraction column than it is to obtain the necessary data and perform a digital computer analysis of the nonlinear effects.

For a fractional-extraction separation, the real value of the linear treatment given here is in providing the preliminary design and economic evaluation and in guiding the pilot-plant experimentation and the final process design. The amount of physical-chemical data required is manageable. A distribution coefficient is measured for each solute component in dilute solution, and a few additional experiments are performed to obtain an estimate of how concentrated the solutions may get before large deviations in the distribution coefficients result. Preliminary designs and economic evaluations may then proceed by the methods presented in this chapter, employing an arbitrary upper limit on the total solute concentration. Pilot-plant experiments on a fractional-extraction column are then used to determine how rapidly the separation deteriorates as the solvent flow rates are decreased, and such data are used to guide the choice of the solvent flow rates for the final process design. Even if it is deemed desirable to obtain the necessary data and develop a computer solution to the nonlinear problem, the linear model will have been quite useful in the early stages by guiding preliminary design considerations.

NOMENCLATURE

$\mathscr{B}$ = molal (or mass) flow rate of a solute component in heavy-solvent stream leaving bottom stage of cascade $\equiv HX_1$

D = distribution coefficient $\equiv Y/X$ at equilibrium

E = extraction factor $\equiv LD/H$

F = stage number of feed stage

$\mathscr{F}$ = molal (or mass) flow rate of solute component fed to stage F, moles/time

G = stage number of second feed stage

$\mathscr{G}$ = molal (or mass) flow rate of solute component fed to stage G, moles/time.

H = molal (or mass) flow rate of heavy solvent, moles/time

$\mathscr{H}$ = molal (or mass) flow rate of a solute component in heavy-solvent feed $\equiv HX_{N+1}$

L = molal (or mass) flow rate of light solvent, moles/time

$\mathscr{L}$ = molal (or mass) flow rate of a solute component in light-solvent feed $\equiv LY_0$

N = number of stages in the cascade; stage number of top stage

R = fractional recovery of solute component

$\mathscr{R}$ = reflux ratio $\equiv$ reflux rate/product rate

$S_{A,B}$ = separating power of the cascade for separating component A from component B; defined by Eq. (3-60)

$\mathscr{T}$ = molal (or mass) flow rate of a solute component in light-solvent stream leaving top stage of cascade $\equiv LY_N$

X = mole (or mass) ratio in heavy phase, moles of solute per mole of heavy solvent (or mass of solute per unit mass of heavy solvent)

Y = mole (or mass) ratio in light phase, moles of solute per mole of light solvent (or mass of solute per unit mass of light solvent)

GREEK LETTERS

$\alpha_{A,B}$ = relative distribution coefficient of component A relative to component B, $\equiv D_A/D_B$
α = same as $\alpha_{A,B}$ but with the understanding that components A and B are defined such that $D_A > D_B$

SUBSCRIPTS

A, B, . . . = solute components A, B . . .
F, G = feed stages numbered F, G
f = feed composition for heavy-solvent feed; but see Eq. (3-25) for significance of Y_f
j = arbitrary stage, numbered j
L = lower section of cascade, below the feed stage
N = top stage in the cascade
$N + 1$ = feed composition for heavy-solvent feed
U = upper section of cascade, above the feed stage
0 = feed composition for light-solvent feed; but see Eq. (3-18) for significance of X_0
1 = bottom stage in the cascade
2, 3, . . . = stages numbered 2, 3, . . . from the bottom

REFERENCES

1. ANON., In Uranium Drama, Chemical Process Plays Star Role, *Chem. Eng.*, 112 (Oct. 1955).
2. G. F. ASSELIN and E. W. COMINGS, *Ind Eng. Chem.*, **42**, 1198 (1950).
3. M. BENEDICT and T. H. PIGFORD, *Nuclear Chemical Engineering*, New York, McGraw-Hill, 1957, Chap. 6.
4. *Ibid.*, Chap. 8.
5. C. D. HARRINGTON and A. E. RUEHLE, *Uranium Production Technology*, New York, Van Nostrand Reinhold, 1959, Chap. 4.
6. A. KREMSER, *Natl. Petroleum News*, **22**, No. 21, 42 (May 21, 1930).
7. E. H. MARTEL, *Chem. Eng.*, 66 (March 29, 1965).
8. W. L. MCCABE and E. W. THIELE, *Ind. Eng. Chem.*, **17**, 605 (1925).
9. U. S. MORELLO and N. POTTENBERGER, *Ind. Eng. Chem.*, **42**, 1021 (1950).

10. S. Peterson and R. G. Wymer, *Chemistry in Nuclear Technology*, Reading, Mass., Addison-Wesley, 1963, Chap. 12.
11. E. G. Scheibel, *Ind. Eng. Chem.*, **47**, 2290 (1955).
12. E. G. Scheibel, Chemicals from Tall Oil by Liquid Extraction, paper presented at Am. Inst. Oil Chemists, Memphis Meeting, 1959.
13. E. G. Scheibel, U. S. Patent 3,177,196 (April 6, 1965).
14. E. G. Scheibel, *Chem. Eng. Progr*, **62**, 76 (1966).
15. T. K. Sherwood and R. L. Pigford, *Absorption and Extraction*, 2nd ed., New York, McGraw-Hill, 1952, Chap. 10.
16. R. E. Treybal, *Liquid Extraction*, New York, McGraw-Hill, 1951, Chaps. 6 and 7.
17. *Ibid.*, Chaps. 9 and 10.
18. *Ibid.*, pp. 347–350.
19. *Ibid.*, pp. 386–388.
20. P. N. Vashist and R. B. Bechkmann, *Ind. Eng. Chem.*, **60**, 43 (1968).
21. J. T. Wood and J. A. Williams, *Trans. Inst. Chem. Engrs.* (*London*), **36**, 382 (1958).

STUDY PROBLEMS

1. Repeat Example Problems 3-1 and 3-2 for cross-flow (instead of countercurrent) cascades with uniform distribution of the light solvent among the stages.
2. One thousand pounds per hour of a mixture of fatty and rosin acids is to be fed to stage 16 of a 31-stage fractional-extraction cascade employing heptane as the light solvent and methyl Cellosolve containing 10 per cent water by volume as the heavy solvent. The mass fractions in the feed and the distribution coefficients in terms of mass ratios for the acids are

Acid	*Mass Fraction in Acid Feed*	*Distribution Coefficient*
Oleic	0.1	4.14
Linoleic	0.5	2.17
Abietic	0.4	1.57

Prepare a graph that shows the effect of the light solvent to heavy solvent flow-rate ratio, L/H, upon the composition and the quantity of both the top and bottom acid product streams.

3. Referring to Problem 2, what value of L/H will produce a symmetrical split between linoleic and abietic acids? At this value of L/H, how many stages would be required to achieve $S_{A,B} = 1.6 \times 10^5$ for the separation of linoleic from abietic acid using a cascade fed at the middle stage? For this cascade, calculate the flow rate of each of the three acids in the top and bottom product streams.
4. Referring to Problem 3, for the cascade design with $S_{A,B} = 1.6 \times 10^5$ calculate the flow rate of each of the three acids in all streams entering and leaving the feed stage. Repeat the calculations for all streams entering and leaving stages 1 and 30. What must be the flow rates of light and heavy solvent if the total acid content of each phase must never exceed 0.07 lb of total acids per pound of solvent?
5. Referring to the feed stream of Problem 2, a cascade is to be designed to recover 90 per cent of the abietic acid from this feed stream at a purity of 99.9 wt % abietic acid. The solubility relationships require that the total acid content of each phase be kept below 0.07 lb of total acids per pound of solvent. Specify values of N, F, L, and H for the cascade design that will accomplish this recovery with the minimum number of stages (without using solute reflux or solvent withdrawal at the feed stage).
6. Repeat Problem 5 using solute reflux at top and bottom reflux ratios of 9:1.
7. Referring to the feed stream of Problem 2, it is desired to separate this feed stream into three product streams, each containing a different acid at a purity of 99.5 wt % pure. How can this be accomplished by liquid–liquid extraction employing the solvents used in Problem 2? Specify the optimum design of this system (but use solvent-free feed streams and use no solute reflux). The limit on total acid content of the phases is to be as in Problem 5. Use the following economic guidelines and cost estimates:
   (a) Pulsed sieve-tray columns will be used for the extraction cascades. Plate efficiency is $33\frac{1}{3}$ per cent—that is, the number of actual sieve plates required is three times the number of ideal stages required.
   (b) The column diameter must be chosen such that the sum of the volumetric flow rates of the heavy and light solvents divided by the column cross-sectional area should be equal to 0.1 ft/sec. This will keep the velocities sufficiently low to permit adequate separation of the phases.
   (c) The installed cost of a sieve-tray column is: cost in dollars per actual plate $= 2.1$ (column diameter in inches)$^{1.57}$
   (d) Light and heavy solvents will be separated from the solutes by boiling and will then be recycled back to the cascade. The installed cost of the boiler plus condenser for a given solvent separation is

$$\text{cost in dollars} = 1.7Q^{0.6}$$

   where Q = boiler heat load in Btu/hr. The cost of heat supplied to the boiler plus cooling in the condenser is \$0.70 per million Btu supplied to the boiler.
   (e) An operating year of 363 days is assumed.
   (f) The total capital investment is estimated to be 2.0 times the installed cost of columns, boilers, and condensers.
   (g) The amortization rate on invested capital is 25 per cent per year.

(h) The operating labor cost will not be considered a variable. Solvent loss will be neglected, as will investment in solvent inventory.

Assume perfect solvent–solute separation in the solvent boilers.

Additional data are

	Density, lb/ft^3	*Heat of Vaporization, Btu/lb*
Light solvent	42.4	137.5
Heavy solvent	60.5	312

8. Derive Equation (3-88).

9. For the optimum design found in Problem 7, consider using solute reflux to see if this will effect a cost reduction. Also consider the possibility of solvent addition at the feed stage.

Binary Distillation 4

Distillation is the most widely used method of separating solutions into essentially pure components on an industrial scale. As in liquid–liquid extraction and many other separation processes, distillation involves the contacting and subsequent separating of two immiscible fluid phases between which the components to be separated distribute in different ratios. But in distillation the two immiscible phases employed are vapor-phase and liquid-phase mixtures of the components to be separated; no additional solvent components are added. Several advantages of this choice of immiscible fluids are immediately apparent. First, the two immiscible fluid phases are readily created, either by partially boiling a liquid stream or by partially condensing a vapor stream. In addition, the components to be separated are not contaminated by the addition of solvent components; therefore, there is no need for employing additional steps to separate solvent and solute components as in liquid–liquid extraction. A third advantage is that the vapor and liquid phases will generally have substantially different densities, except when operating very near the critical point, and the interfacial tension is usually relatively large; therefore, the separation of the two phases by gravity is usually accomplished quite simply. Finally, both phases are fluids and therefore easily handled and conveyed. Indeed, while pumps and gravity flow are employed for the liquid phase, pumps and blowers are generally not needed for conveying the vapor phase because relatively small increments in boiling temperature can produce vapor pressure differences sufficient to cause the vapor streams to flow through the pipes and other equipment. These advantages make distillation the most economical means of accomplishing the majority of the purification steps in the chemical process industries.

Of course, distillation processes have limitations that rule out distillation as a purification method in certain instances. The temperature of operation must be between the triple-point and the critical-point temperatures for the components being separated. In some separation problems, this requirement dictates a temperature range either too hot or too cold for convenient and economical operation. The pressure of the distillation operation is dictated by the vapor pressures of the components to be separated at the operating temperature. For some systems, the pressure might be quite high at convenient operating temperatures, and this may be a disadvantage. More usual, however, is the situation in which the components to be separated have very low volatilities; thus the pressure level will be quite low unless high operating temperatures are employed. When, in addition, the components to be separated are thermally unstable at high temperatures, distillation can be accomplished only under high-vacuum conditions in order that the operating temperature be kept acceptably low. In vacuum operation the vapor density is low, and the equipment size for a given mass throughput of vapor is large, resulting in a relatively high cost of the distillation operation. Therefore, distillation is often ruled out for separating mixtures of materials which are quite nonvolatile and also heat sensitive. In cases of this type, liquid–liquid extraction is often a more attractive alternative. Of course, the ease of separating one component from another by distillation is dependent upon their vapor–liquid equilibrium behavior. In some instances, the separation cannot be achieved by distillation because of an unfavorable vapor–liquid equilibrium relationship.

VAPOR–LIQUID EQUILIBRIUM

When a liquid phase and a vapor phase containing the same two components are brought into intimate contact, the two components will diffuse from one phase to the other until the liquid- and vapor-phase compositions correspond to equilibrium. It is conventional to express the liquid- and vapor-phase compositions as mole fractions, denoting liquid-phase mole fractions by the symbol x and the vapor-phase mole fractions by the symbol y. For a given temperature, if a series of equilibrium measurements is made corresponding to different equilibrium-liquid-phase compositions, it is found that for each liquid-phase composition there will generally be a different vapor-phase composition and a different system pressure at equilibrium.

Figure 4-1 presents isothermal vapor–liquid equilibrium data for the benzene–toluene system. The upper pair of curves represents equilibrium at 110.6°C, which is the normal boiling point of toluene. The uppermost curve is a plot of the equilibrium pressure versus x_b, the liquid-phase mole fraction of benzene. This curve is called the *bubble-point pressure curve*. Beneath this curve lies the plot of *dew-point pressures* at 110.6°C, a graph of the equilibrium pressure versus the equilibrium-vapor-phase mole fraction of benzene. The intersections of these two curves with a horizontal line correspond to the values of x_b, y_b, and the pressure, P, which can be in equilibrium at 110.6°C. For example, consider a liquid-phase mixture of benzene and toluene with a benzene mole fraction of 0.3. For this liquid-phase composition, the bubble-point pressure is 1.4 atm. At this same pressure of 1.4 atm, it is seen from the dew-point pressure curve at 110.6°C that the equilibrium-vapor composition corresponds to a benzene mole fraction of 0.5. Thus, at a pressure of 1.4 atm, corresponding to the dashed horizontal line in Fig. 4-1, the equilibrium-liquid composition is read from the upper curve

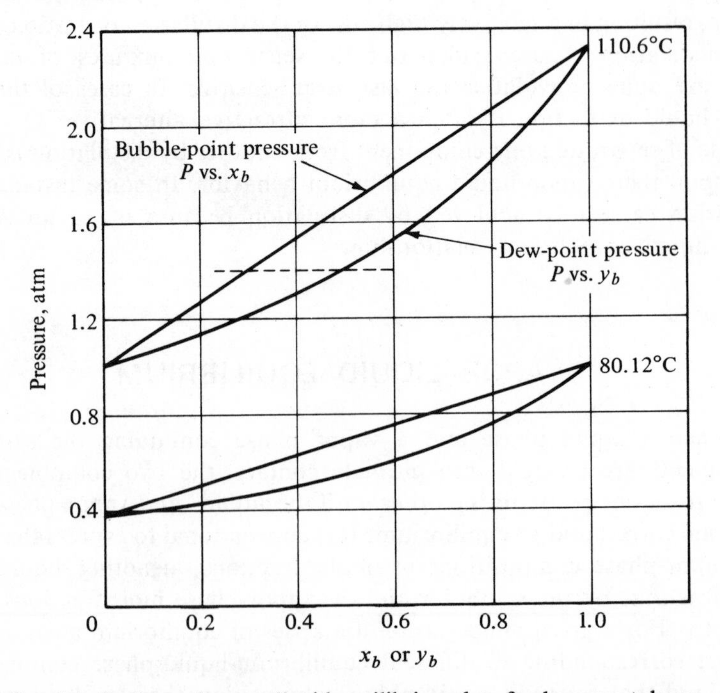

Fig. 4-1. Isothermal vapor–liquid equilibrium data for benzene–toluene.

as 0.3 mole fraction benzene and the equilibrium-vapor composition is read from the lower curve as 0.5 mole fraction benzene.

If a liquid-phase mixture of benzene and toluene with a benzene mole fraction of 0.3 is held at a pressure greater than 1.4 atm at a temperature of 110.6°C, the liquid is said to be a *subcooled* liquid, and this liquid cannot be in equilibrium with any vapor composition at that temperature and pressure. If the pressure is reduced to 1.4 atm, the liquid is right at its boiling point, or *bubble point*, and an additional small reduction in pressure will cause some of the liquid to vaporize. The equilibrium vapor will be richer in benzene than is the liquid; the first vapor will have a benzene mole fraction of 0.5.

If a vapor mixture of benzene and toluene with a benzene mole fraction of 0.5 is held at a pressure of less than 1.4 atm at a temperature of 110.6°C, the vapor is said to be *superheated*, and it cannot be in equilibrium with a benzene–toluene liquid phase of any composition at that temperature and pressure. If the pressure is raised to 1.4 atm, the vapor is said to be at its *dew point*, and an additional small increase in pressure will result in the condensation of some of the vapor. The first equilibrium liquid condensed out will have a benzene mole fraction of 0.3 and will, therefore, be richer in toluene than is the vapor phase.

The lower pair of curves in Fig. 4-1 presents the bubble-point and dew-point curves at a constant temperature of 80.12°C, which is the normal boiling point of benzene. As the mole fractions approach unity, the bubble-point and the dew-point pressure curves both approach a pressure of 1 atm, which is the vapor pressure of benzene at 80.12°C. As the mole fractions approach zero, the curves approach a pressure of 0.38 atm, which is the vapor pressure of toluene at this temperature. In contrast, the uppermost curves, corresponding to 110.6°C, approach a pressure of 1 atm, the vapor pressure of toluene at this temperature, as the mole fractions approach zero. As the mole fractions approach unity, the curves approach a pressure of 2.33 atm, the vapor pressure of benzene at 110.6°C. At temperatures higher than 110.6°C, the isothermal equilibrium curves lie above those shown in Fig. 4-1, and at temperatures less than 80.12°C, the curves lie below those shown in Fig. 4-1.

Figure 4-2 presents isobaric vapor–liquid equilibrium data for the benzene–toluene system. The uppermost pair of curves presents the equilibrium data for a constant pressure of 2 atm. The uppermost curve is a plot of the temperature versus the vapor-phase mole fraction of benzene; this is called the *dew-point temperature curve*. The curve beneath this is a graph of the temperature versus the liquid-phase mole fraction of benzene; it is called the *bubble-point temperature curve*. A horizontal line, such as the dashed line shown in Fig. 4-2, intersects these two curves at the points corresponding to values of T, x_b, and y_b which correspond to equilibrium. Thus a liquid mixture of benzene and toluene with a benzene mole fraction of 0.65 at a pressure of 2 atm is a subcooled liquid at temperatures less than 113.4°C.

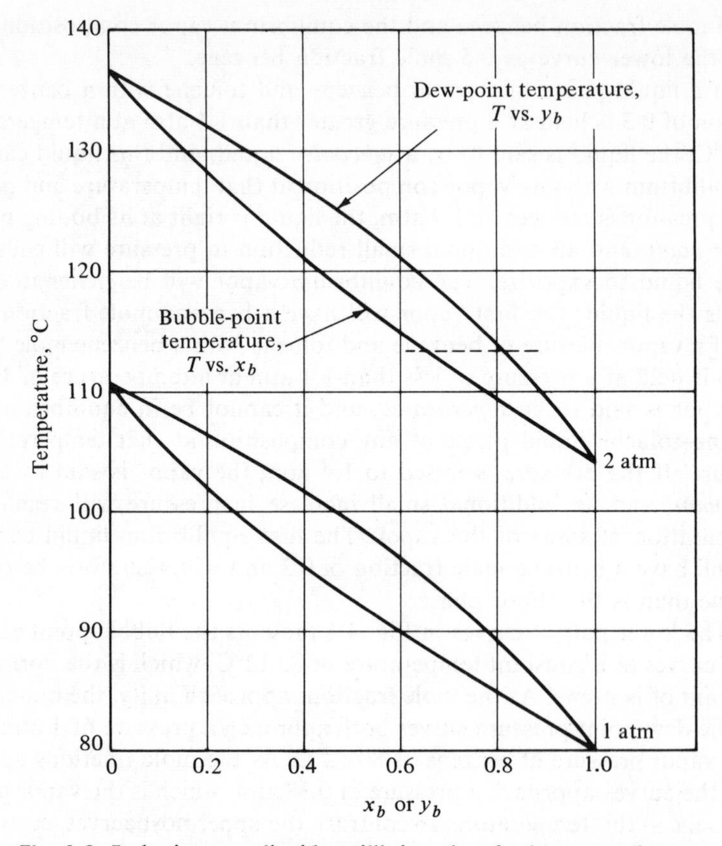

Fig. 4-2. Isobaric vapor–liquid equilibrium data for benzene–toluene.

At that temperature the liquid is a saturated liquid at its boiling point, and the equilibrium vapor has a benzene mole fraction of 0.81. A vapor-phase mixture of benzene and toluene with a benzene mole fraction of 0.81 at a pressure of 2 atm is a superheated vapor at temperatures greater than 113.4°C. At that temperature, its dew-point temperature, it is in equilibrium with a liquid that is 65 mole per cent benzene.

The lower pair of curves in Fig. 4-2 represents the isobaric equilibrium data at a pressure of 1 atm. As the mole fractions approach unity, both the dew-point and the bubble-point curves approach a temperature of 80.12°C, the normal boiling point of benzene. As the mole fractions approach zero, both curves approach 110.6°C, the normal boiling point of toluene. Similarly, the uppermost curves in Fig. 4-2 approach the boiling points of benzene and toluene at a pressure of 2 atm when the mole fractions approach unity and zero, respectively. Isobaric equilibrium curves for pressures greater than 2 atm would lie above the curves shown in Fig. 4-2, and curves for pressures less than 1 atm would lie below the curves shown in Fig. 4-2; but the shapes of those curves would be similar.

Generally speaking, isobaric equilibrium data are more useful than isothermal equilibrium data for the design of distillation columns, because many distillation columns operate at approximately a constant pressure. While the bubble-point and dew-point temperature curves shown in Fig. 4-2 are a complete representation of the isobaric vapor–liquid equilibrium data, the representation of Fig. 4-3 is often more convenient for distillation-column design. In Fig. 4-3 the equilibrium-vapor mole fraction of benzene is plotted versus the liquid-phase benzene mole fraction. Each point corresponds to a different temperature, as can be seen in Fig. 4-2, but the temperature variation is not indicated in Fig. 4-3. Thus the y–x diagram of Fig. 4-3 is not a complete representation, because it does not include the equilibrium temperature for each point on the curve. But, as will be seen subsequently, the equilibrium vapor and liquid compositions are of greatest concern in distillation-column design, and the y–x diagram is a very convenient representation of the equilibrium-composition relationship. The

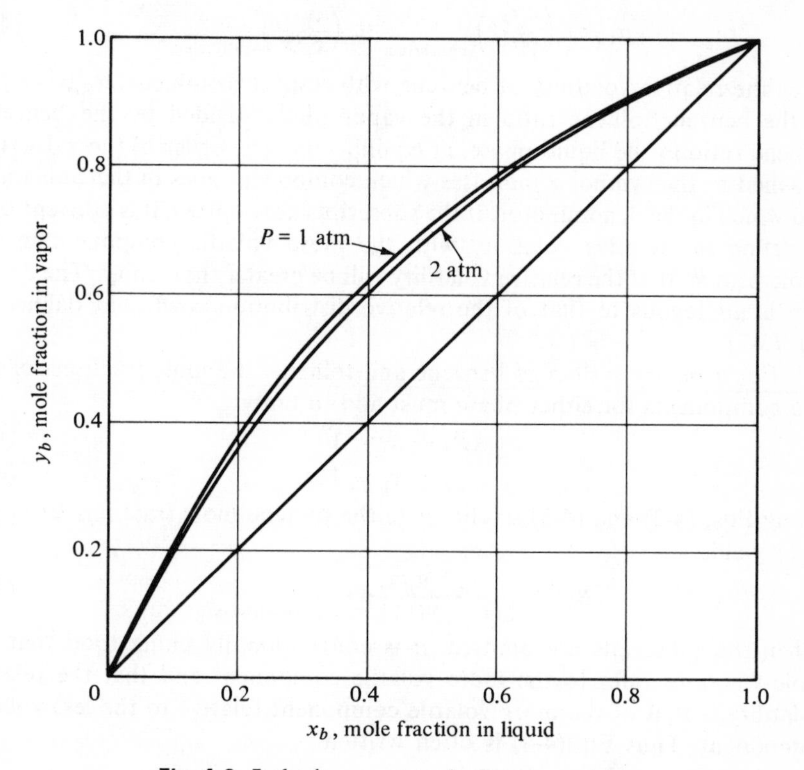

Fig. 4-3. Isobaric y–x curves for benzene–toluene.

upper curve in Fig. 4-3 is the y–x diagram representing the isobaric equilibrium data at 1 atm. The lower curve represents the isobaric equilibrium data at a pressure of 2 atm. It is seen in Fig. 4-3 that the y–x curve changes but little when the pressure changes from 1 to 2 atm. Of course, the bubble-point and dew-point temperatures change substantially, as indicated in Fig. 4-2, but the relationship between the equilibrium compositions of the liquid and vapor phases is relatively insensitive to the system pressure.

The equilibrium curves in Fig. 4-3 lie above the 45° diagonal line, which is shown for reference. Thus benzene is said to be more volatile than toluene because it distributes preferentially into the vapor phase while toluene distributes preferentially into the liquid phase. Of course, if the mole fractions of toluene in the vapor phase and in the liquid phase had been plotted in Fig. 4-3, the equilibrium curve would lie beneath the 45° diagonal line. Generally speaking, it is conventional to plot the mole fractions of the more volatile species when expressing binary vapor–liquid equilibrium data; thus the equilibrium curve will lie above the 45° diagonal line.

The extent to which benzene is more volatile than toluene is measured quantitatively by the *relative volatility*, defined as

$$\alpha_{b,t} \equiv \left(\frac{y_b/x_b}{y_t/x_t}\right)_{\text{equilibrium}} \equiv \left(\frac{y_b/y_t}{x_b/x_t}\right)_{\text{equilibrium}} \tag{4-1}$$

Thus the relative volatility of benzene with respect to toluene, $\alpha_{b,t}$, is equal to the benzene/toluene ratio in the vapor phase divided by the benzene/toluene ratio in the liquid phase, at equilibrium. The order of the subscripts attached to the symbol α indicates which component goes in the numerator and which in the denominator. If the subscripts are omitted, it is conventional to define the relative volatility with the more volatile component in the numerator so that the relative volatility will be greater than unity. The definition is analogous to that of the relative distribution coefficient defined in Eq. (3-2).

For a binary system of benzene and toluene, the mole fractions of the two components for either phase must add to unity:

$$x_b + x_t = 1 \tag{4-2}$$

$$y_b + y_t = 1 \tag{4-3}$$

Using Eqs. (4-2) and (4-3) to eliminate the toluene mole fractions from Eq. (4-1) yields

$$\alpha_{b,t} \equiv \left[\frac{y_b/x_b}{(1 - y_b)/(1 - x_b)}\right]_{\text{equilibrium}} \tag{4-4}$$

When the subscripts are omitted, it is conventionally understood that the mole fractions refer to the more volatile component and that the relative volatility is that of the more volatile component relative to the less volatile component. Thus Eq. (4-4) is often written

$$\alpha \equiv \left[\frac{y/x}{(1 - y)/(1 - x)}\right]_{\text{equilibrium}} \tag{4-5}$$

with the understanding that y and x refer to the mole fractions of the more volatile component.

Reading the uppermost curve in Fig. 4-3, y_b is found to be 0.374, 0.711, and 0.912 when x_b is 0.2, 0.5, and 0.8, respectively. Substitution of these values into Eq. (4-4) yields values of 2.39, 2.46, and 2.59 for the relative volatility at the three respective compositions. Thus it is seen that the relative volatility of benzene with respect to toluene at a pressure of 1 atm lies in the range (approximately) 2.4 to 2.6 for liquid compositions between 20 and 80 mole per cent benzene. For the higher pressure of 2 atm the relative volatility lies in the range (approximately) 2.2 to 2.4. Thus it is seen that for the benzene–toluene system the relative volatility does not vary greatly at a given pressure. As the pressure increases, the relative volatility decreases slightly.

If the relative volatility can be considered constant, the y–x equilibrium curve can be represented by the expression

$$y = \frac{\alpha x}{1 + (\alpha - 1)x} \tag{4-6}$$

which is obtained by simply rearranging Eq. (4-5). Figure 4-4 presents a graphical representation of Eq. (4-6) for several different values of α. The

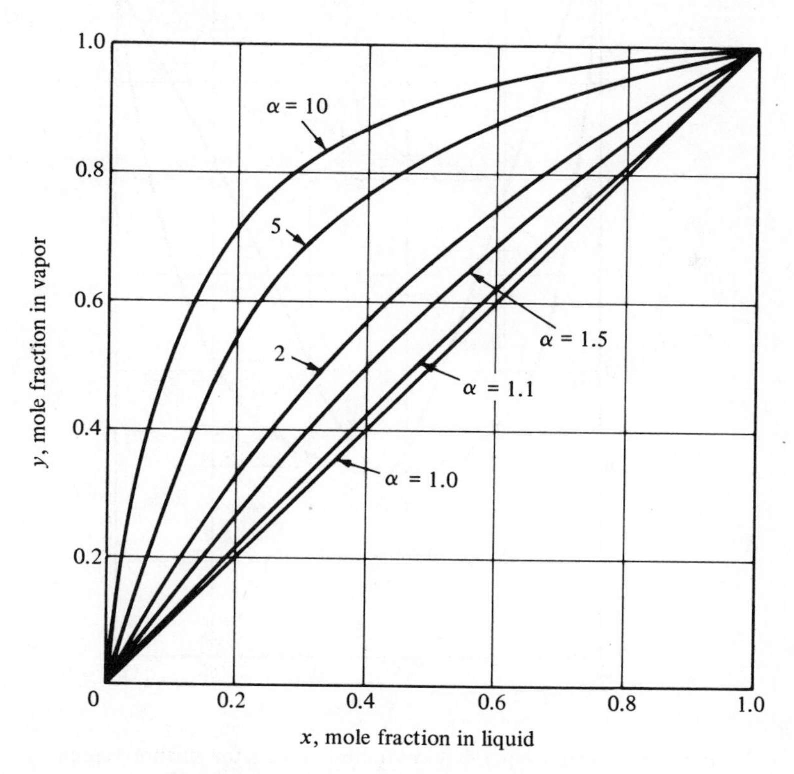

Fig. 4-4. y–x curves for different values of α.

lowest curve is the 45° line, $y = x$, which corresponds to Eq. (4-6) with a relative volatility equal to unity. The other curves, for relative volatility values greater than unity, all lie above the 45° diagonal line to an extent that increases as α increases. Thus the relative volatility of a system is a measure of the extent to which the vapor–liquid equilibrium curve departs from the 45° diagonal line, and it is, therefore, a measure of the degree of separation between the two components that can be obtained in a single equilibrium contact between a vapor phase and a liquid phase containing these components.

4-1. Azeotropes

The benzene–toluene system is a system of normal relative volatility, because the relative volatility of benzene with respect to toluene is greater

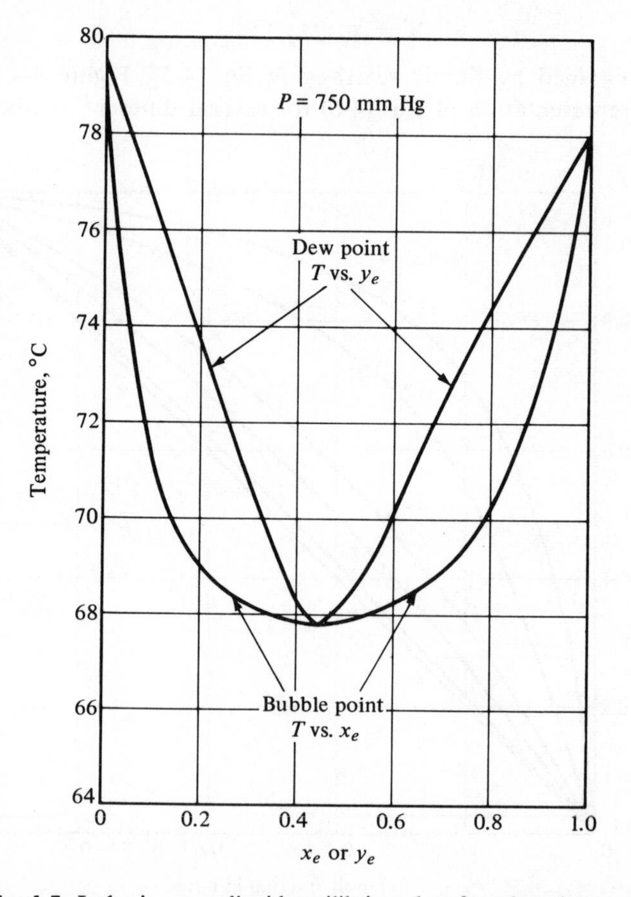

Fig. 4-5. Isobaric vapor–liquid equilibrium data for ethanol–benzene.

than unity over the entire range of liquid-phase composition. In contrast to this behavior is that of the ethanol–benzene system, which forms a minimum-boiling azeotrope. Figure 4-5 presents isobaric equilibrium data for the ethanol–benzene system at a pressure of 750 mm of mercury. In contrast to the curves shown in Fig. 4-2, the dew-point and the bubble-point curves are not single-valued curves lying between the boiling point of one pure component and the boiling point of the other pure component. Instead, the curves dip below the boiling points of both components down to a minimum temperature of 67.8°C, the azeotropic temperature. For values of x_e less than 0.445, ethanol is more volatile than benzene, the value of y_e being greater than the value of x_e with which it is in equilibrium. But at a value of x_e equal to 0.445, the bubble-point curve and the dew-point curve come back together, and the vapor-phase composition is equal to the liquid-phase composition; this is the azeotropic composition. For values of x_e greater than 0.445, y_e is seen to be less than the value of x_e with which it is in equilibrium; thus ethanol is less volatile than benzene in this composition range.

Figure 4-6 presents the y–x representation of the ethanol–benzene

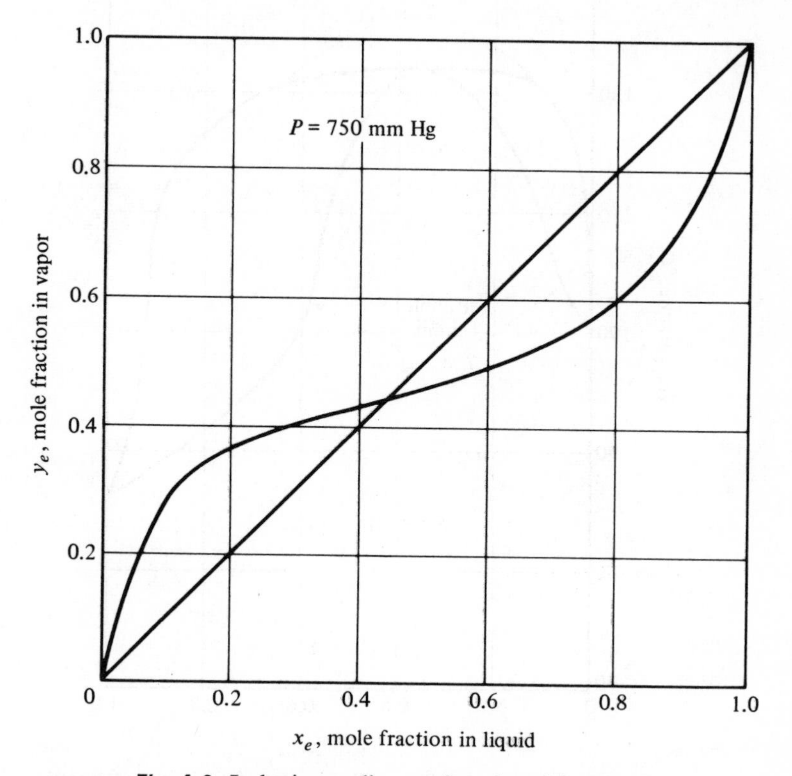

Fig. 4-6. Isobaric y–x diagram for ethanol–benzene.

isobaric equilibrium data at a pressure of 750 mm of mercury. Instead of the curve lying completely above the 45° diagonal line, as in Figs. 4-3 and 4-4, the equilibrium curve lies above the 45° diagonal line at low values of x_e and below the 45° diagonal line at high values of x_e. The equilibrium curve intersects the 45° diagonal line at a value of $x_e = 0.445$, the azeotropic composition. At this intersection point the equilibrium curve has a slope less than unity. (This would still be the case if y_b were plotted versus x_b; the qualitative shape of the curve would not be changed.) The intersection of the curve with the 45° diagonal line with a slope less than unity is characteristic of minimum-boiling azeotropic systems, in which the azeotropic temperature is lower than the boiling point of either pure component. The relative volatility of ethanol with respect to benzene is greater than unity for values of x_e less than 0.445, and it is less than unity for values of x_e greater than 0.445. At the azeotropic composition the relative volatility is equal to unity.

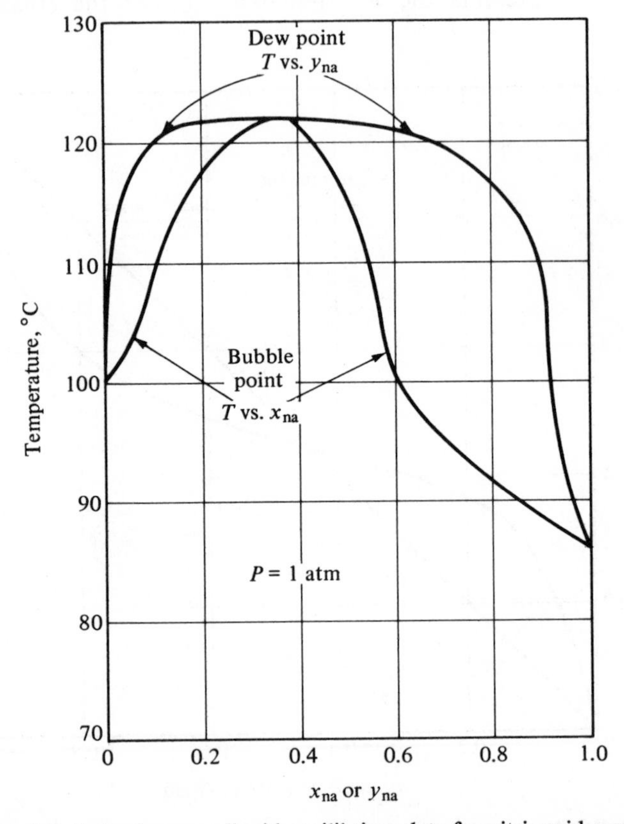

Fig. 4-7. Isobaric vapor–liquid equilibrium data for nitric acid–water.

In contrast to the benzene–toluene system, in which the relative volatility varies but little with composition, the relative volatility of ethanol with respect to benzene varies from a value greater than 4 at low values of x_e to a value less than $\frac{1}{4}$ at high values of x_e.

Figures 4-7 and 4-8 present isobaric equilibrium data for the nitric acid–water system at a pressure of 1 atm. This system exhibits a maximum-boiling azeotrope with a composition of 38.3 mole per cent nitric acid and an azeotropic temperature of 121.9°C. The boiling points of the pure components are 86°C for nitric acid and 100°C for water, and it can be seen in Fig. 4-7 that the bubble-point and dew-point curves rise substantially above the boiling point of either pure component up to the azeotropic temperature of 121.9°C. For values of x_{na} less than 0.383, y_{na} is seen to be less than the value of x_{na} with which it is in equilibrium; thus nitric acid is less volatile than water in this range. For values of x_{na} greater than 0.383, y_{na} is greater than x_{na}; thus nitric acid is more volatile than water. This behavior is also seen in Fig. 4-8, where the y–x curve is seen to cross the 45° diagonal line at the azeotropic composition of 38.3 mole per cent nitric acid. At this point

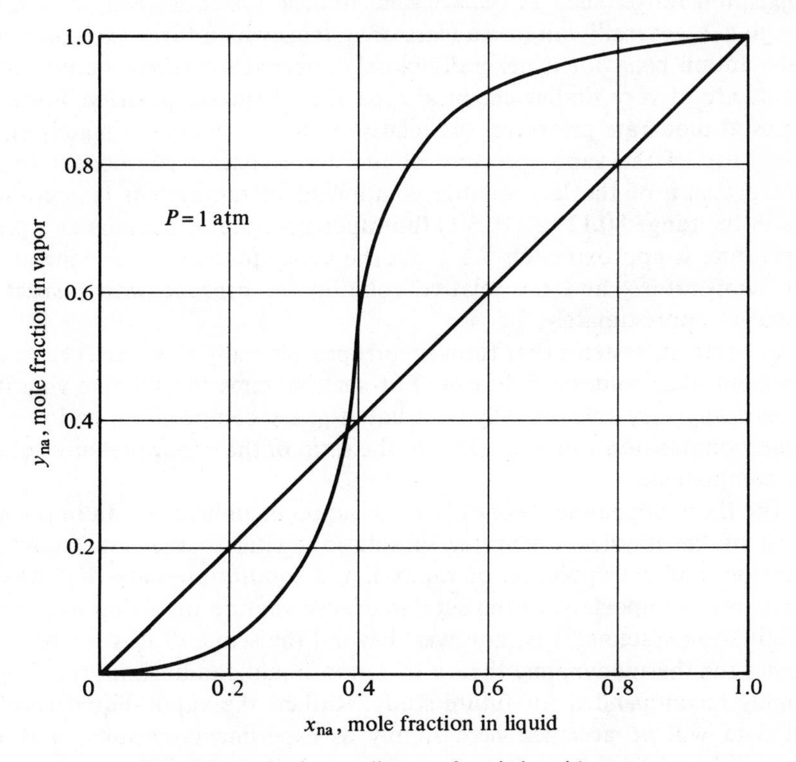

Fig. 4-8. Isobaric y–x diagram for nitric acid–water.

the equilibrium curve intersects the 45° diagonal line with a slope greater than unity. (The qualitative shape of the curve would be the same and the slope would be greater than unity if y_w were plotted versus x_w instead of plotting y_{na} versus x_{na}.) This slope greater than unity at the intersection point is characteristic of maximum-boiling azeotropes, in which the azeotropic temperature is greater than the boiling point of either pure component.

Since the relative volatility for an azeotropic system crosses unity at the azeotropic composition, no equilibrium separation between the two components is obtained at the azeotropic point. Thus in systems that form azeotropes it is generally not possible to separate the components by distillation if the desired separation involves crossing the azeotropic composition.

4-2. Ideal and nonideal solutions

Systems for which the relative volatility varies but little over the entire composition range, such as the benzene–toluene system shown in Figs. 4-1 through 4-3, generally fall into a class of systems which form *ideal solutions*. Ideal-solution behavior is generally closely approached when the two components are of very similar chemical type and of similar polarity. For such systems at moderate pressures, the relative volatility is approximately equal to the ratio of the vapor pressure of the more volatile component to the vapor pressure of the less volatile component at the system temperature. Thus in the range 80.12 to 110.6°C the vapor pressure of benzene at a given temperature is approximately 2.5 times the vapor pressure of toluene at the same temperature, and the relative volatility of benzene with respect to toluene is approximately 2.5.

In contrast, systems that form azeotropes generally show marked deviations from ideal-solution behavior. For such systems the relative volatility will generally vary substantially with liquid-phase composition, and only at a single composition will it be equal to the ratio of the vapor pressures of the pure components.

The thermodynamic theory of vapor–liquid equilibrium is an important branch of the physical chemistry of solutions vital to the understanding, prediction, and extrapolation of vapor–liquid equilibrium data. Knowledge in this area is important to the success of any venture involving the design of distillation systems. It is, however, beyond the scope of this textbook to consider the thermodynamic theory of vapor–liquid equilibrium; the subject is highly recommended for future study. Rather, the vapor–liquid equilibrium data will be accepted here simply as experimental results, and the focus will be upon the distillation-cascade design and performance.

FRACTIONAL DISTILLATION

Continuous distillation involves the countercurrent contacting of a vapor stream with a liquid stream in a series of contacting stages to effect a separation between components of different volatilities. If a binary mixture is to be distilled into relatively pure components, a fractional-distillation cascade is employed in which the feed stream is introduced into the middle of the cascade and relatively pure components are withdrawn from the top and bottom of the cascade.

Figure 4-9 shows a diagram of a fractional-distillation column. An upward-flowing vapor stream and a downward-flowing liquid stream are

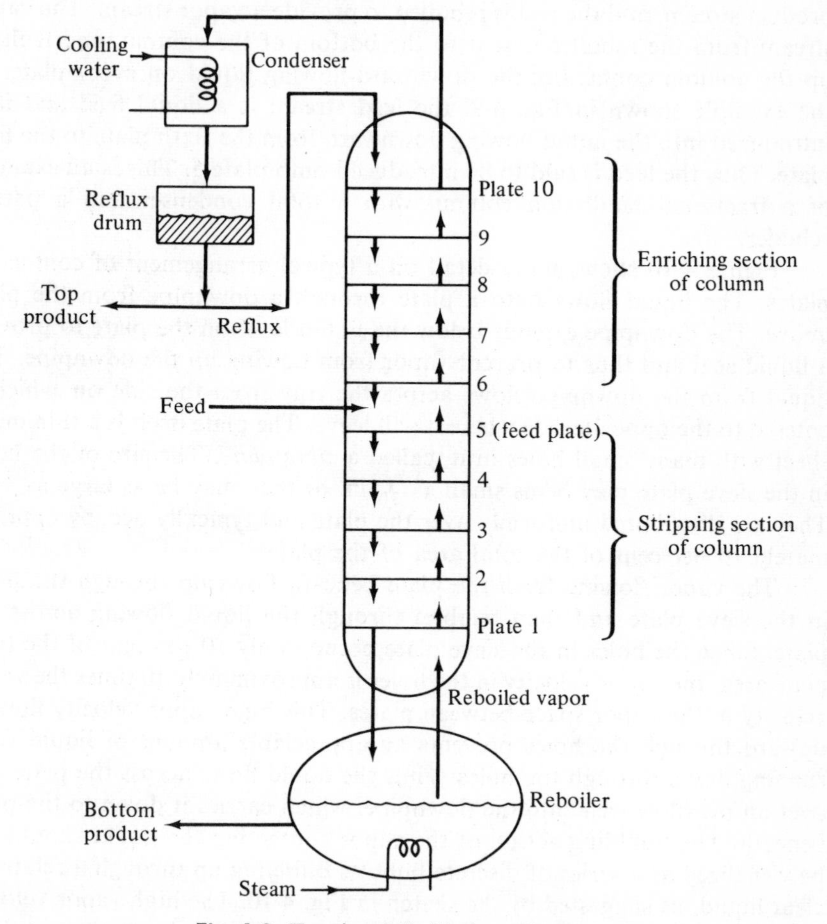

Fig. 4-9. Fractional-distillation column.

brought into contact on a series of contacting plates within a vertical cylindrical column. Figure 4-9 shows a column with 10 contacting plates. The vapor leaving the top plate goes to a condenser where it is condensed, for example, by contacting a water-cooled metal coil. The condensate is collected in a storage tank called the reflux drum, and part of the liquid leaving this drum is returned to the column as a reflux liquid stream while the rest of the liquid leaving the reflux drum is withdrawn as the top, or *distillate*, product stream. The reflux liquid stream is added to the top of the column, and it flows down onto successive plates in the column, contacting the upward-flowing vapor stream on every plate. The liquid flowing from the bottom plate is sent to a reboiler where it is partially boiled, for example, by contact with a steam-heated metal coil. The liquid stream entering the reboiler is only partially reboiled; a portion of it is withdrawn as a bottom liquid product stream, and the rest is reboiled to provide a vapor stream. The vapor stream from the reboiler is sent to the bottom of the column, and it flows up the column contacting the downward-flowing liquid on every plate. In the example shown in Fig. 4-9, the feed stream is a liquid feed and it is introduced into the liquid flowing downward from the sixth plate to the fifth plate. Thus the feed is said to be introduced onto plate 5. This is an example of a fractional-distillation column with a total condenser and a partial reboiler.

Figure 4-10 shows more detail on a typical arrangement of contacting plates. The liquid flows onto a plate through a downpipe from the plate above. The downpipe extends below the liquid level on the plate to provide a liquid seal and thus to prevent vapor from flowing up the downpipe. The liquid from the downpipe flows across the tray from the side on which it entered to the opposite side where it will leave. The plate itself is a thin metal sheet with many small holes in it, called a *sieve plate*. The size of the holes in the sieve plate may be as small as $\frac{1}{30}$ in. or they may be as large as 1 in. They are distributed uniformly over the plate and typically occupy approximately 10 per cent of the total area of the plate.

The vapor flowing from the plate beneath flows up through the holes in the sieve plate and then bubbles through the liquid flowing across the plate. Since the holes in the sieve plate occupy only 10 per cent of the total plate area, the vapor velocity in the holes is approximately 10 times the vapor velocity in the vapor space between plates. This high vapor velocity flowing upward through the holes prevents an appreciable amount of liquid from running down through the holes. Thus the liquid flows across the plate and over an overflow weir into the downpipe, which carries it down to the plate beneath. The bubbling action of the vapor contacting the liquid should not be visualized as a series of discrete bubbles bubbling up through a relatively clear liquid, as suggested by the sketch in Fig. 4-10. The high vapor velocity in the holes and the high vapor loading typically used in distillation columns

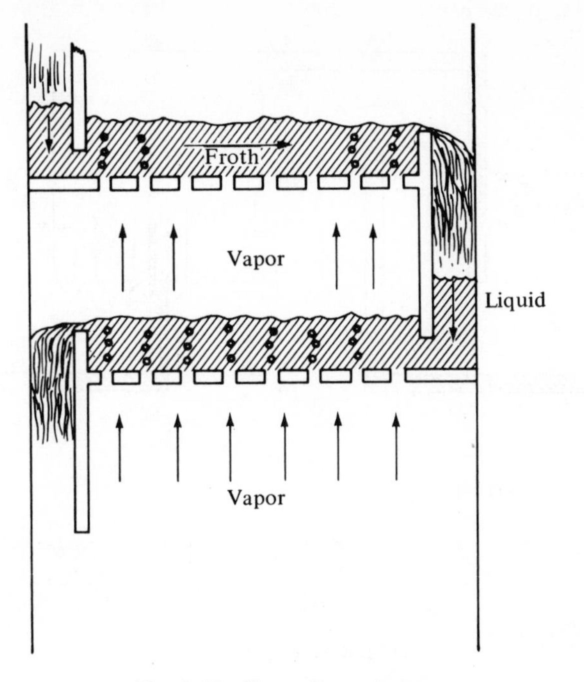

Fig. 4-10. Sieve-plate column.

results in violent agitation which produces a froth of liquid and vapor bubbles on top of the plate. The froth is approximately 50 per cent liquid by volume, and the many small vapor bubbles provide an abundance of gas–liquid interfacial area, which is required for intimate contacting of the phases.

The sieve plate is an inexpensive and effective method of contacting the liquid stream and the vapor stream in a distillation column, and it is the preferred method of construction in many distillation systems today. A major disadvantage of the sieve plate is that it cannot be operated satisfactorily over a wide range of vapor rates. If the liquid and vapor rates in the column are reduced too much, the vapor velocity up through the holes in the sieve plate will cease to be high enough to prevent liquid from draining down through the holes, or *weeping*. When a substantial percentage of the liquid does drain down through the holes instead of flowing across the tray to the outlet weir, bypassing of the vapor stream by the liquid stream results, and the liquid and vapor streams are not contacted effectively.

One means of preventing the liquid from flowing down through the vapor holes, and of thus ensuring positive control over the liquid flow across the plate, is the use of a *bubble cap*, which is sketched in Fig. 4-11. Instead of a simple sieve plate, the plate is fitted with bubble caps. Each hole

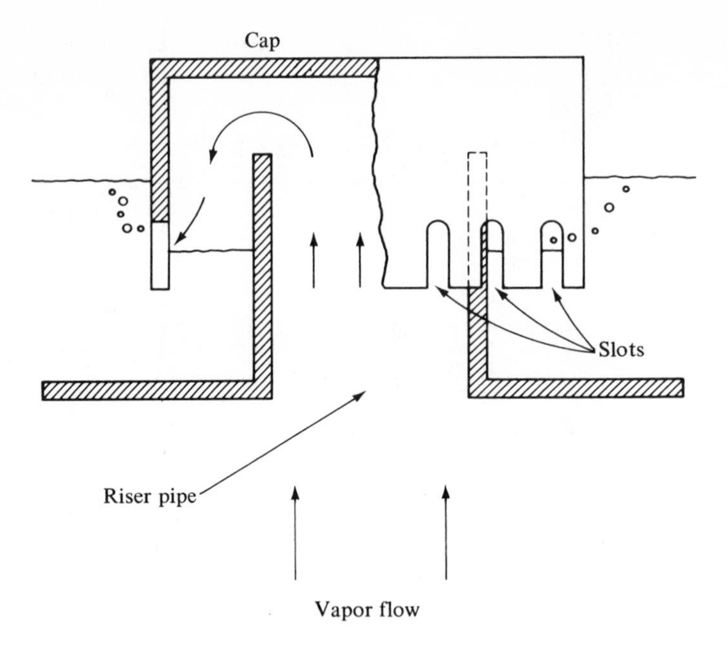

Fig. 4-11. Bubble cap.

in the plate through which the vapor flows is fitted with a riser pipe, which extends higher than the level of the froth on the plate. Over each riser pipe is fitted a cap which has a number of slots extending up from its lower edge. The vapor from the plate beneath flows up through the riser pipe and then downward between the cap and the riser pipe and finally bubbles into the liquid through the slots at the bottom of the cap. If the vapor flow is reduced or even stopped, the liquid cannot flow down through the vapor holes because the height of the riser pipe is greater than the liquid depth upon the plate. The use of bubble caps is an effective means of extending the range of vapor rates over which the column can be successfully operated, but it is by no means the only method employed. Various other means, such as valve trays with check-valve types of arrangements, are also employed. Generally speaking, the bubbling actions and thus the contacting efficiencies of these devices are similar to that of the sieve tray. The construction of these trays is usually somewhat more costly than that of the simple sieve plates, but the added flexibility of operation can justify the additional cost in some situations.

The distillation column shown in Fig. 4-9 is a fractional-distillation column with the feed introduced in the middle of the cascade. The plates below the feed point are referred to as the *stripping section* of the column and are required for the effective removal of the more volatile component from the bottom product. The plates above the feed point are referred to as the *enriching section* of the column and are necessary for the effective removal of the less volatile component from the top product. When both the top and the bottom products must be relatively pure, both sections of the column are necessary. This is not always the case, however, and simple stripping columns, with the feed introduced at the top, and also enriching columns, with the feed introduced at the bottom, are sometimes employed when only one of the product streams must be of high purity.

BINARY DISTILLATION-COLUMN DESIGN

The mathematical model employed for distillation-column design and for the simulation of distillation-column performance is based upon material and energy balances around each plate in the column, equilibrium relationships between the vapor and liquid compositions, and a measure of the efficiency of vapor–liquid contacting and thus the degree of approach to equilibrium achieved by the contacting plates. A detailed analysis of some of these considerations can be avoided by making certain simplifying assumptions which are widely used and acceptably accurate in many cases. These simplifying assumptions will be reviewed, and the mathematical model of a distillation column will be developed employing these simplifications. Later, the simplifying assumptions will be examined more closely and some of the restrictions will be removed.

4-3. Steady-state operation

Many distillation columns are operated continuously and their operation is approximately steady with time. A feed stream of approximately constant composition is fed at a steady rate to the column. Vapor is boiled at approximately a steady rate in the reboiler and is condensed at a steady rate in the condenser. Likewise, top and bottom product streams are withdrawn at approximately constant rates, and liquid reflux is returned to the top of the column at a constant rate. After an initial transient start-up period, liquid and vapor compositions at every point within the column become approximately constant with time, as do the compositions of the product streams. Of course, all distillation systems are not operated in this manner, but the majority are, and this justifies a focus on the analysis of steady-state operation. A steady-state mathematical model assumes perfectly steady operation in which the flow rate and composition of any stream at any point within the system are absolutely constant with time. This assumption simplifies the material-balance equations to the point of simply requiring that the flow rate of a given component in all streams entering any section of the column must be equal to the flow rate of that component in the streams leaving that section. The energy balance is also simplified in a similar manner.

4-4. Constant molal overflow

A detailed consideration of the energy balance can be avoided by the assumption of constant molal overflow. With this simplification, it is assumed that the molal flow rate of the upward-flowing vapor stream and the molal flow rate of the downward-flowing liquid stream are both constant throughout any section of the column to which no feed stream is added and from which no product stream is withdrawn. For example, referring to Fig. 4-9, the molal flow rate of vapor from plate 4 up to plate 5 would be assumed equal to the molal flow rate of vapor from the reboiler to plate 1. The compositions of these two vapor streams would be different, but their flow rates would be assumed to be equal when expressed in units of total moles flowing per unit time. Likewise, the molal flow rates of the vapor streams leaving plates 1, 2, and 3 would also be equal to the molal flow rate of vapor from the reboiler. Similarly, the molal flow rate of liquid leaving plates 2, 3, 4, and 5 would all be assumed to be equal to the molal flow rate of liquid flowing from plate 1 to the reboiler. Similarly, the molal vapor rate leaving plates 6, 7, 8, and 9 would all be assumed to be equal to the molal rate of flow of vapor from plate 10 to the condenser, and the molal liquid rates leaving plates 6, 7, 8, 9, and 10 would all be assumed to be equal to the liquid-reflux rate to the top plate. Thus, for a column such as that shown in Fig. 4-9 which has a single feed stream and which has no side-product streams withdrawn from the column itself, the liquid and vapor rates are constant throughout the stripping section and throughout the enriching section.

The introduction of the feed stream will generally cause the liquid flow rate in the stripping section to be different from that in the enriching section and will also generally cause the vapor flow rate in the enriching section to be different from that in the stripping section. If the feed stream is a liquid at its bubble-point temperature, it is usually assumed that the molal vapor rate in the enriching section of the column is equal to that in the stripping section, but that the molal liquid rate in the stripping section is greater than the molal liquid rate in the enriching section by an amount equal to the molal flow rate of the feed stream. On the other hand, if the feed stream is a vapor at its dew-point temperature, it is assumed that the molal liquid-overflow rates in the stripping and enriching sections are equal but that the molal vapor rate in the enriching section is greater than the molal vapor rate in the stripping section by an amount equal to the molal flow rate of the vapor feed stream.

If the feed stream is a mixture of liquid and vapor, its introduction is assumed to increase the liquid rate in the stripping section over that in the enriching section by an amount equal to the rate of flow of the liquid fraction of the feed stream, and also to increase the vapor rate in the enriching section over that in the stripping section by an amount equal to the vapor fraction of the feed-stream flow rate. If the feed stream is a liquid below its bubble-point temperature, this subcooled liquid will actually condense some vapor; this results in a molal liquid rate in the stripping section greater than the molal liquid rate in the enriching section by an amount which exceeds the molal flow rate of the feed stream. In this case, the molal vapor rate in the enriching section will be less than the molal vapor rate in the stripping section by an amount equal to the rate of vapor condensation by the cold liquid feed stream. Similarly, if the feed stream is a superheated vapor, the molal liquid-overflow rate in the stripping section will be less than that in the enriching section, and the vapor rate in the enriching section will exceed that in the stripping section by an amount greater than the molal feed rate. In this case, the superheated vapor stream results in the vaporization of some of the liquid overflow to the feed plate.

Of course, the vapor and liquid rates in any section of the column could be altered arbitrarily by adding or removing heat to or from that section of the column. In general, it is not desirable to do this purposefully, but heat will transfer between the column and its surroundings if the column operates above or below "room temperature." But most industrial distillation systems employ columns of relatively large diameter and thus of relatively small surface area per unit volume. Therefore, the rate of heat transfer between the column and its surroundings is usually quite small. Furthermore, columns that operate substantially above or below room temperature are usually insulated to minimize heat exchange with the surroundings. Thus most industrial distillation columns operate approximately adiabatically.

For an adiabatic column the assumption of constant molal overflow is equivalent to assuming that 1 mole of vapor of one composition has an amount of energy of vaporization equivalent to 1 mole of vapor of another composition and, therefore, that the condensation of 1 mole of vapor of one composition releases just the right amount of energy to boil up 1 mole of vapor of any other composition. The accuracy of this approximation depends upon the chemical system. The assumption of constant molal overflow is surprisingly accurate for some systems, whereas for other systems its accuracy may not be acceptable. It is a widely used assumption, and it greatly simplifies distillation-column design. The assumption of constant molal overflow will be adopted for the development here but will be examined in greater detail and modified later.

4-5. Theoretical plates

The concept of a distillation column containing *theoretical plates*, or *ideal plates*, is often employed in distillation-column design. A theoretical plate is one that is assumed to contact the vapor and liquid streams so as to bring them to equilibrium with each other. Thus the composition of the vapor stream *leaving* the plate and the composition of the liquid stream *leaving* the plate are assumed to correspond to the equilibrium relationship. Employing the concept of a column composed of theoretical plates may in some instances be equivalent to assuming that the plates in a distillation column will indeed bring the liquid and vapor phases approximately to equilibrium. In many other instances, however, it is known that this is not the case, but the theoretical-plate concept is still very useful as a means of breaking down the overall design problem into two smaller problems, the cascade design on the one hand and the considerations of contacting efficiency on the other. Thus, although the ultimate design question must be that of determining how many actual plates are required to effect a given separation, the designer will often address himself first to the question of how many theoretical plates are required, and then to the question of how many actual plates are equivalent to a given number of theoretical plates. Similarly, in approaching a *simulation* problem, one may calculate the effect of a change in some operating variable, such as the reboil rate, upon the purity of the products produced by a column containing a given number of theoretical plates and learn most of the important aspects of the way in which a real column would respond to that change in operation.

The simplification of considering a column composed of theoretical plates will be adopted here at the beginning and will be maintained for most of the discussion of distillation-column design and performance. Some consideration will be given to the concepts of an overall column efficiency and a plate efficiency, which measure the departure of the real plates from ideal behavior, later on.

4-6. Isobaric operation

The liquid flows down a distillation column under the influence of gravity, but a pressure gradient is required to cause the vapor to flow up the column. Therefore, the pressure is highest in the reboiler and lowest in the condenser, and the pressure decreases with distance up the column. Most of the pressure drop occurs because of the high-velocity vapor flow through the holes in the plates and because of the liquid head on the plates. As mentioned earlier, it is not necessary to use pumps or blowers to supply the extra pressure to the vapor stream. A small increase in the reboiler temperature will cause an increase in vapor pressure sufficient to cause the vapor to flow up the column.

The pressure drop in a distillation column is typically less than 0.01 atm per actual plate, and thus a 50-plate column would usually have a pressure drop of less than 0.5 atm. For a column with a condenser pressure of

1 atm or higher, the *percentage* change in the pressure throughout the column is therefore usually moderate. Since the y–x equilibrium diagram is generally quite insensitive to moderate pressure changes, as shown in Fig. 4-3, the y–x curve applying to the bottom of the column will not be very different from that applying to the top of the column. It is common, therefore, to neglect the pressure variation within the column and to use the y–x diagram corresponding to the average pressure, at least for the initial design.

For distillation columns operating at high vacuum, the column pressure drop can be a *large percentage* of the absolute pressure in the condenser, and the effect of the pressure variation throughout the column may be important. This will complicate the column design, because it will require that the effect of pressure on the equilibrium y–x diagram be accounted for.

The prediction of pressure drop across a distillation plate involves *fluid mechanics*; this subject will not be treated here. Rather, the column will be assumed to be isobaric in this treatment. The fluid mechanics of liquid and vapor flow in a distillation column will be left for future study.

4-7. Material balances

The assumptions of constant molal overflow and theoretical plates simplify the mathematical description of a distillation column to the point where only material balances and vapor–liquid equilibrium relationships need be considered. In the component material balances, the flow rate of a given component in a given flowing stream is expressed as the product of the total flow rate of that stream and of the concentration of the component in question in that stream. Since it is the *molal* flow rates of the liquid and vapor streams in the column that are assumed to be constant, it is convenient to express the component flow as the product of the molal flow rate of the stream and the component mole fraction in that stream. This is especially convenient because vapor–liquid equilibrium data are usually expressed in terms of mole fractions, this custom having its origin in the thermodynamic theory of vapor–liquid equilibrium and carrying through to the presentation of experimental results quite generally.

The overall material balance for the entire column requires that the total moles flowing into the column must equal the total moles flowing out of the column per unit time:

$$F = D + B \tag{4-7}$$

where F, B, and D represent the molal flow rates of the feed stream, the bottom product stream, and the distillate or top product stream, respectively. The component material balance for the entire column is written

$$Fz_F = Dx_D + Bx_B \tag{4-8}$$

where x_D and x_B represent the mole fractions of the component in the distillate and bottom products, respectively, and z_F represents the mole fraction of the component in the feed stream. If the feed stream is a mixture of liquid and vapor, z_F is the average mole fraction in the mixture. An equation of the form of Eq. (4-8) could be written for each component, but in a binary system only one such equation is necessary, because for each stream the mole fractions of the two components add to unity; thus the additional equation is simply the difference between Eq. (4-7) and Eq. (4-8) written for one of the components. Equation (4-8) can be written for either of the two components in a binary mixture, but it is conventional to write it for the more volatile component and, in the absence of a subscript denoting which component is referred to, it is conventionally assumed that the mole fractions refer to the more volatile component (in an azeotropic system it should be specified which component is referred to in order to avoid ambiguity).

Instead of writing material balances around each plate in the column, it is more convenient to write the material balance around a section of the column that includes the liquid and vapor streams within the column at a given point and also the top or bottom product stream. For example, for the bottom section of the column, the material-balance envelope indicated by the dashed line in Fig. 4-12 is chosen. The envelope slices through the column between plate m and plate $m + 1$ and also slices through the bottom product exit line. Thus the only stream entering this material-balance envelope is the liquid-overflow stream from plate $m + 1$ to plate m, and the two streams leaving are the vapor stream rising from plate m to plate $m + 1$ and the bottom product stream leaving the system. The liquid-overflow rate and the flow rate of the rising vapor stream in the bottom section of the column are denoted L' and V', respectively, and the subscripts on the mole fractions, y and x, refer to the number of the plate from which the stream in question is *leaving*. With this notation, the material balances for the dashed envelope at the bottom of the column in Fig. 4-12 are

$$L' = V' + B \tag{4-9}$$

$$L'x_{m+1} = V'y_m + Bx_B \qquad (\text{for } m = 0, 1, \ldots, f - 1) \tag{4-10}$$

for the total material balance and the component material balance, respectively. Equation (4-10) relates the compositions of the upward-flowing vapor stream and the downward-flowing liquid stream at any point within the stripping section of the column. For $m = 0$, it relates the composition of the vapor leaving the reboiler to the composition of the liquid leaving the bottom plate. For $m = 1$, it relates the composition of the vapor leaving the first plate to the composition of the liquid flowing down from the second plate. The equation applies for values of m up to $f - 1$, at which point it relates the composition of the vapor rising from the plate beneath the feed plate to the composition of the liquid flowing down from the feed plate, plate f.

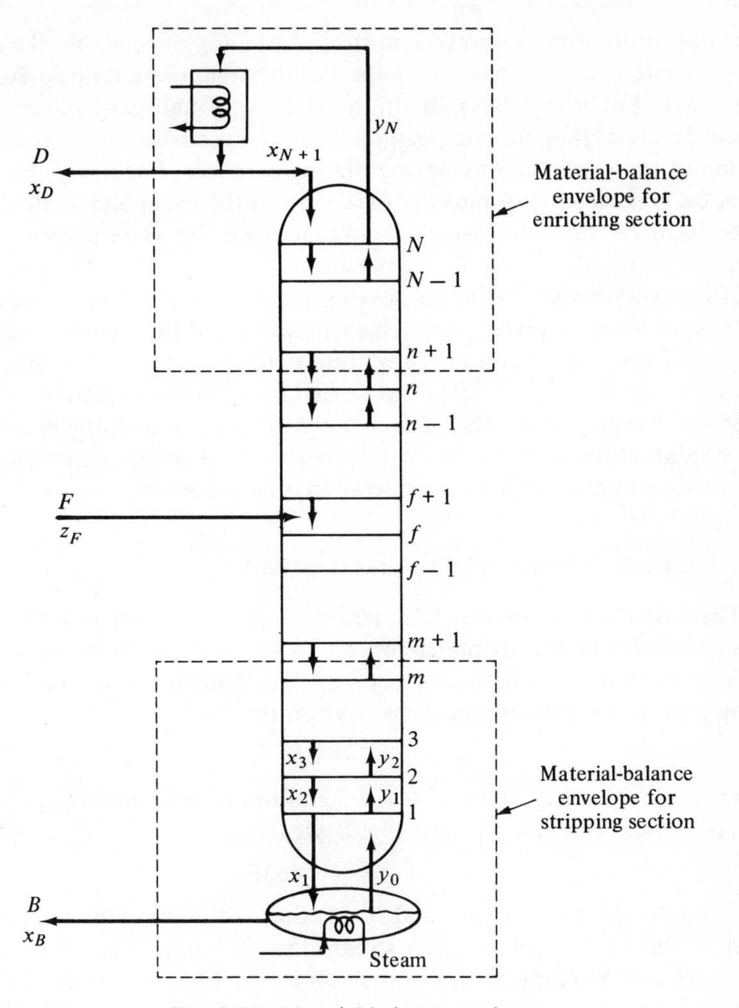

Fig. 4-12. Material-balance envelopes.

For the upper section of the column, the material-balance envelope is drawn to include the rising vapor stream, the liquid-overflow stream, and the distillate-product stream, as shown by the dashed line at the top of the column in Fig. 4-12. Total and component material balances for this material-balance envelope are

$$V = L + D \tag{4-11}$$

$$Vy_n = Lx_{n+1} + Dx_D \qquad (\text{for } n = f, f+1, \ldots, N) \tag{4-12}$$

where L and V represent the molal flow rates of the liquid and the vapor, respectively, in the enriching section of the column. Equation (4-12) relates the composition of the vapor stream rising from any plate within the enriching section to the composition of the liquid stream overflowing from the plate above. For $n = N$, this equation is simply a material balance around the condenser, relating the composition of the reflux stream, x_{N+1}, to the composition of the vapor stream leaving the top plate, y_N. For $n = f$, Eq. (4-12) relates the composition of the vapor rising from the feed plate, plate f, to the composition of the liquid overflow stream from the plate above the feed plate.

It is assumed that the vapor leaving plate f has the same composition as the vapor entering plate $f + 1$. This will be valid if this vapor stream does not contact the feed stream in rising from plate f to plate $f + 1$. If the feed stream is a liquid feed, it is often introduced into the downpipe which carries the liquid overflow from plate $f + 1$ down to plate f, and the assumption will be valid. Other methods of introducing the feed stream and their effects upon this assumption will be considered in a later section.

4-8. Thermal condition of the feed stream

The introduction of the feed stream will generally cause the liquid and vapor rates in the stripping section to be different from those in the enriching section. The difference between the liquid rates in the stripping section and in the enriching section is given by

$$L' - L = qF \tag{4-13}$$

This equation is best thought of as the definition of the quantity q. Combining Eqs. (4-7), (4-9), (4-11), and (4-13) yields

$$V - V' = (1 - q)F \tag{4-14}$$

It is apparent from Eqs. (4-13) and (4-14) that the quantity q may be interpreted as the fraction of the feed stream that is liquid. Thus the quantity qF is the liquid content of the feed stream, and this joins with the liquid overflow from the enriching section to provide the liquid overflow in the stripping section. Similarly, the quantity $(1 - q)F$ is the vapor content of the feed stream, and this increases the vapor flow rate in the enriching section above that in the stripping section. Thus, when the feed stream is a mixture of liquid and vapor, q is simply the liquid fraction in the feed, expressed as a mole fraction. For the special case of a feed stream that is a liquid at its bubble point, $q = 1$; for a vapor at its dew point, $q = 0$. When the feed stream is a subcooled liquid, q is greater than unity, and it is given approximately by

$$q = 1 + \frac{c_{PL}(T_{BP} - T_F)}{\Delta H} \tag{4-15}$$

The numerator in the second term on the right-hand side of Eq. (4-15) represents the heat required to raise 1 mole of the subcooled liquid from the temperature at which the feed is introduced into the column, T_F, to the bubble-point temperature of the feed liquid at the column pressure, T_{BP}. The denominator of this term is the molal heat of vaporization of the feed liquid. For a superheated vapor feed, q is negative, and it is given approximately by

$$q = -\frac{c_{PV}(T_F - T_{DP})}{\Delta H} \tag{4-16}$$

The numerator on the right-hand side of Eq. (4-16) represents the heat required to raise 1 mole of vapor from its dew-point temperature, T_{DP}, to the temperature at which it is introduced into the column, T_F. Again the denominator represents the heat of vaporization of 1 mole of feed. Equations (4-15) and (4-16) could be generalized by noting that q is equal to the heat that must be *added* to 1 mole of feed to bring it to a saturated vapor divided by the molal heat of vaporization.

4-9. Plate-to-plate calculation

The material-balance equations, together with the equilibrium relationship, can be used for a plate-to-plate calculation to determine the number of theoretical plates required to effect a given separation with a specified reflux ratio (or reboil ratio). First, the external flow rates and compositions are determined by the overall material-balance equations, Eqs. (4-7) and (4-8). For a specified feed rate or an arbitrary basis such as 100 moles of feed, and for a specified feed composition, Eqs. (4-7) and (4-8) are solved simultaneously to yield the distillate- and bottom-product flow rates for specified distillate and bottom product compositions, or they may be solved for the compositions if the flow rates are specified. Next, the internal flow rates of vapor and liquid within the column are computed for a specified reflux ratio, L/D, or for a specified vapor rate expressed as the reboil ratio, V'/B, or as the vapor-to-feed ratio, V'/F. For a specified value of q, Eqs. (4-9), (4-11), (4-13), and (4-14) determine the liquid and vapor rates within the column.

The plate-to-plate calculation is then carried forward using Eqs. (4-10) and (4-12), together with the equilibrium relationship. Starting at the bottom of the column, for example, the composition of the vapor leaving the reboiler, y_0, is commonly assumed to be in equilibrium with the bottom product liquid composition. This assumption involves the concept of an ideal reboiler and certain ideas about mixing within the reboiler itself, and this assumption will be examined in more detail later. For now it will simply be assumed that the reboiler is operated in such a manner as to produce a vapor stream and a liquid bottom product stream which are in equilibrium with each other. Using the value of y_0 thus obtained, Eq. (4-10) is used to calculate x_1, the composition of the liquid flowing from the bottom tray. The equilibrium curve then yields the value y_1 in equilibrium with x_1. This value of y_1 is then used in Eq. (4-10) for the calculation of x_2. The equilibrium relationship then yields y_2 in equilibrium with x_2. Equation (4-10) is used again to compute x_3 from this value of y_2. This procedure is repeated, using alternately the equilibrium relationship to give the value y_m in equilibrium with x_m, followed by the use of Eq. (4-10) to compute x_{m+1} from that value of y_m.

Equation (4-10) applies for m values only up to $f - 1$, thus yielding a value of x_f. The value of y_f is determined as that value in equilibrium with x_f, but then x_{f+1} is determined by Eq. (4-12), not by Eq. (4-10). Therefore, when the feed is introduced, thus defining plate f, Eq. (4-12), the material balance in the enriching section of the column, must be used to compute the liquid compositions for plates above the feed plate. The choice of the feed-plate location is somewhat arbitrary, but in most cases it is desired to locate the feed plate such that the total number of equilibrium plates required to effect the separation will be a minimum. To accomplish this in the plate-to-plate calculation, one may compute the liquid composition from the next plate above both by Eq. (4-10) and Eq. (4-12). If the resulting value computed from Eq. (4-10) is greater (when calculating up the column) than the value computed from Eq. (4-12), then the value from Eq. (4-10) is employed and introduction of the feed is delayed until a higher plate. As the calculation proceeds up the column, the first time the value of x computed by Eq. (4-12) exceeds the value computed by Eq. (4-10), then the value from Eq. (4-12) is employed, implying that the plate beneath was the choice for the feed plate. The calculation then proceeds up to the top employing Eq. (4-12), the material balance for the enriching section of the column.

Assuming that the condenser at the top of the column is a total condenser (one that completely condenses all the vapor sent to it) the composition of the distillate product, x_D, and the composition of the liquid reflux to the top plate, x_{N+1}, will both be equal to the composition of the vapor rising from the top plate, y_N. Therefore, the plate-to-plate calculation is terminated when the composition of the vapor rising from a plate is determined to be equal to (or just greater than) the specified distillate product composition. The number of that plate then corresponds to the number of theoretical plates required to effect the specified separation.

The plate-to-plate calculation will be illustrated with the following example problems.

EXAMPLE PROBLEM 4-1 A saturated-liquid mixture containing 60 mole per cent benzene and 40 mole per cent toluene is to be distilled continuously into a distillate product containing 90 mole per cent benzene and a bottom product containing 5 mole per cent benzene. The fractional-distillation column will operate at approximately a constant pressure of 1 atm. The reflux ratio, L/D, will be 2.0. How many theoretical plates must the column have if the feed is introduced onto the eighth plate from the bottom?

Solution: Since no feed rate was specified, an arbitrary basis of 100 moles of feed will be chosen. This is equivalent to expressing all flow rates in units of moles per unit time where the time unit is chosen to make the feed rate 100. The distillate and bottom product compositions were specified, but the flow rates of these product streams must be determined from the simultaneous solution of Eqs. (4-7) and (4-8). Equation (4-7) becomes

$$100 = D + B$$

With the understanding that all mole fractions represent the mole fraction of benzene, Eq. (4-8) becomes the benzene material balance:

$$60 = 0.9D + 0.05B$$

Solving these two equations simultaneously yields

$$D = 64.7, \qquad B = 35.3$$

With a reflux ratio of 2.0, the liquid reflux rate is equal to twice the distillate product rate, or

$$L = (2)(64.7) = 129.4$$

The vapor rate in the enriching section of the column is given by Eq. (4-11):

$$V = 129.4 + 64.7 = 194.1$$

Since the feed is a saturated liquid, $q = 1$, and Eq. (4-13) becomes

$$L' = 129.4 + 100 = 229.4$$

Equation (4-14) then gives the vapor rate in the bottom of the column,

$$V' = V = 194.1$$

Equation (4-9) is not independent of these other equations, and it can be seen that the vaues computed for L', V', and B do satisfy Eq. (4-9).

With these values, Eq. (4-10) becomes

$$x_{m+1} = 0.846y_m + 0.00769$$

which represents the material balance for the stripping section. The plate-to-plate calculation will be carried up the column, beginning with the specified bottom product composition,

$$x_B = 0.05$$

The composition of the vapor leaving the reboiler is assumed to be in equilibrium with the bottom product composition. It can be read directly from the equilibrium curve, which is the upper curve in Fig. 4-3. The value read is

$$y_0 = 0.108$$

The material balance for the stripping section is then used to compute the composition of the liquid flowing from the bottom plate:

$$x_1 = (0.846)(0.108) + 0.00769 = 0.0991$$

Reading the equilibrium curve at this value of x_1, the value of y_1 is read as

$$y_1 = 0.202$$

The stripping-section material balance then yields

$$x_2 = (0.846)(0.202) + 0.00769 = 0.1787$$

Alternating use of the equilibrium curve and the stripping-section material balance then yields the following values:

$$y_2 = 0.34$$

$$x_3 = (0.846)(0.34) + 0.00769 = 0.269$$

$$y_3 = 0.505$$

$$x_4 = (0.846)(0.505) + 0.00769 = 0.435$$

$$y_4 = 0.655$$

$$x_5 = (0.846)(0.655) + 0.00769 = 0.562$$

$$y_5 = 0.762$$

$$x_6 = (0.846)(0.762) + 0.00769 = 0.653$$

$$y_6 = 0.823$$

$$x_7 = (0.846)(0.823) + 0.00769 = 0.704$$

$$y_7 = 0.857$$

$$x_8 = (0.846)(0.857) + 0.00769 = 0.733$$

$$y_8 = 0.873$$

Since the feed is introduced onto the eighth theoretical plate, the value of x_9 is computed from y_8 with the material balance for the enriching section, which is given by Eq. (4-12). Employing the flow rates calculated earlier, Eq. (4-12) becomes

$$x_{n+1} = 1.5y_n - 0.45$$

which is the enriching-section material balance. Inserting $y_8 = 0.873$ into this material balance gives

$$x_9 = (1.5)(0.873) - 0.45 = 0.86$$

From the equilibrium curve it follows that

$$y_9 = 0.94$$

Thus the computed composition of the vapor rising from the ninth theoretical plate exceeds the required benzene mole fraction in the distillate product; therefore, nine theoretical plates are more than adequate to effect the desired separation. Since y_8 was only 0.873, eight theoretical plates are not sufficient in this case.

EXAMPLE PROBLEM 4-2 Repeat Example Problem 4-1 but with the requirement that the feed will not necessarily be introduced onto the eighth plate but rather will be introduced onto whatever plate is necessary in order that the total number of theoretical plates in the column will be minimized.

Solution: Referring to Example Problem 4-1, assume that the feed is introduced onto the third plate. In this case, x_4 is computed from y_3 using the enriching-section material balance, not the stripping-section material balance. Thus the value of x_4 would be

$$x_4 = (1.5)(0.505) - 0.45 = 0.3075$$

This value is smaller than the value of x_4 calculated in Example Problem 4-1. Therefore, introducing the feed onto the third plate would not be advantageous. Using the enriching-section material balance to compute x_5 and x_6 from the values of y_4 and y_5 obtained in the solution to Example Problem 4-1 yields

$$x_5 = (1.5)(0.655) - 0.45 = 0.532$$

$$x_6 = (1.5)(0.762) - 0.45 = 0.693$$

This value of x_5 is smaller than the corresponding value computed in Example Problem 4-1, but this value of $x_6 = 0.693$ is greater than the value of x_6 computed with the stripping-section material balance in Example Problem 4-1. Therefore, it is advantageous to introduce the feed onto the fifth theoretical plate in order that x_6 will be computed with the enriching-section material balance rather than with the stripping-section material balance. Using $x_6 = 0.693$, the equilibrium curve yields

$$y_6 = 0.85$$

Again using the enriching-section material balance

$$x_7 = (1.5)(0.85) - 0.45 = 0.825$$

Reading the equilibrium curve at $x_7 = 0.825$ gives

$$y_7 = 0.923$$

This value of y_7 exceeds the benzene mole fraction required in the distillate product; therefore, seven theoretical plates are more than adequate to effect this separation when the feed stream is introduced onto the fifth theoretical plate. Since y_6 was less than the required distillate product composition, six theoretical plates are not sufficient.

McCABE–THIELE DIAGRAM

The plate-to-plate calculation of a binary fractional-distillation column, described in the previous section, is most readily visualized on a McCabe–Thiele diagram [4]. Figure 4-13 shows a McCabe–Thiele diagram representing the solution of Example Problem 4-2. The equation representing the stripping-section material balance is plotted as a straight line in Fig. 4-13. This line is commonly called the stripping-section *operating line* or the lower operating line. Similarly, the enriching-section material-balance equation is plotted as a straight line commonly called the enriching-section operating line or simply the upper operating line. The specified bottom product compo-

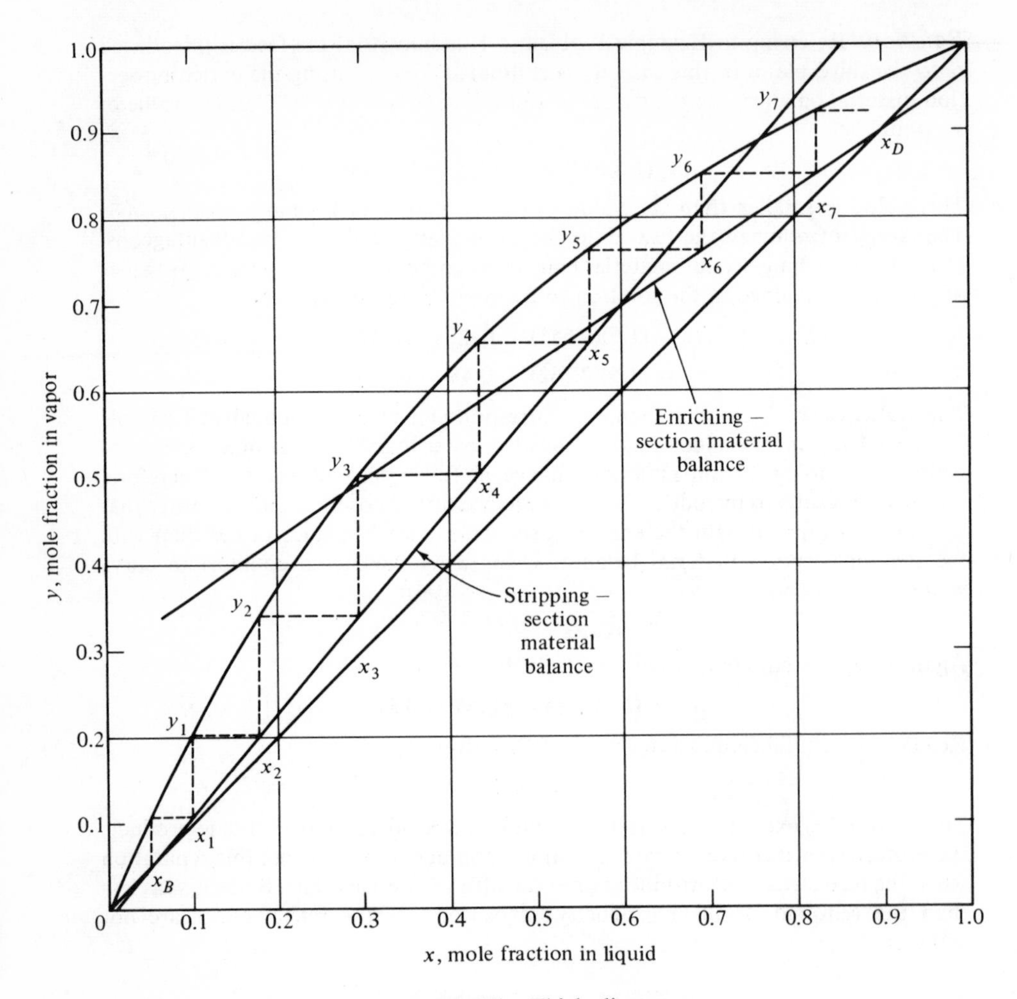

Fig. 4-13. McCabe–Thiele diagram.

sition is $x_B = 0.05$. At this value on the equilibrium curve the value of y_0 is located. A horizontal step to the right over to the lower operating line gives the value of x_1. A vertical step up to the equilibrium curve then gives the value of y_1, and another horizontal step to the right back to the lower operating line gives the value of x_2. Continuing up the column, a step up to the equilibrium curve and then over to the right back to the operating line advances the composition from the x value for one plate up to the x value for the plate above. The point (y_4, x_5) lies on the lower operating line, but the point (y_5, x_6) lies on the upper operating line; therefore, the feed is introduced onto the fifth theoretical plate. The value of y_7 exceeds 0.9, the specified distillate product composition; therefore, seven theoretical plates will produce a somewhat better separation than that which is required.

The McCabe–Thiele diagram can be used to accomplish graphically the plate-to-plate design of a binary distillation column. But it is often more convenient to carry out the actual calculations arithmetically, especially when a digital computer is used. The great advantage of the McCabe–Thiele diagram is that it provides a graphical means of visualizing the distillation-column design and performance, and especially the way in which the design or performance of a distillation column varies with variations in the independent parameters, such as the reflux ratio, the product purities, the thermal condition of the feed, and the relative volatility of the system. This visualization is enhanced substantially by focusing upon the general properties of the McCabe–Thiele diagram, especially the locations of the operating lines and how they vary with the independent parameters of interest.

4-10. Slopes of the operating lines

The upper operating line in a McCabe–Thiele diagram is a graph of Eq. (4-12) plotted as y_n versus x_{n+1}. The plot is represented by a straight line with a slope L/V, the ratio of the liquid flow rate to the vapor flow rate in the enriching section of the column. In terms of the reflux ratio, defined as

$$\mathscr{R} \equiv \frac{L}{D} \tag{4-17}$$

it is clear from Eq. (4-11) that the slope of the upper operating line is also equal to

$$\frac{L}{V} = \frac{\mathscr{R}}{\mathscr{R} + 1} \tag{4-18}$$

The lower operating line on a McCabe–Thiele diagram is a graphical representation of Eq. (4-10) plotted as y_m versus x_{m+1}. This plot is a straight line with a slope L'/V', the ratio of the liquid flow rate to the vapor flow rate in the stripping section of the column. Equation (4-9) can be rearranged to the form

$$\frac{L'}{V'} = 1 + \frac{1}{V'/B} \tag{4-19}$$

in which the slope of the lower operating line is expressed in terms of the reboil ratio, V'/B.

When the reflux ratio and the reboil ratio are high, the liquid and vapor flow rates within the distillation column are large relative to the feed and product rates, and it is seen from Eqs. (4-18) and (4-19) that the slopes of the two operating lines approach unity. The slope of the upper operating line is always less than unity but approaches unity from beneath as the reflux ratio approaches infinity, as seen from Eq. (4-18). The slope of the lower operating line is always greater than unity, but it approaches unity from above as the reboil ratio approaches infinity, as seen from Eq. (4-19). At very low reflux ratios, the slope of the upper operating line approaches zero, while the slope of the lower operating line approaches infinity at very low reboil ratios.

4-11. Intersection with the diagonal

The equation of the upper operating line on a McCabe–Thiele diagram is, according to Eq. (4-12),

$$y = \frac{L}{V}x + \frac{Dx_D}{V} \tag{4-20}$$

This straight line intersects the 45° diagonal line, represented by

$$y = x \tag{4-21}$$

at values of x and y corresponding to the distillate product composition. The generality of this statement can be demonstrated by solving Eqs. (4-20) and (4-21) simultaneously for the intersection point, giving

$$x_i = y_i = \frac{Dx_D/V}{1 - L/V} = x_D \tag{4-22}$$

The second half of Eq. (4-22) follows from Eq. (4-11).

In a similar manner, the intersection of the lower operating line,

$$y = \frac{L'}{V'}x - \frac{Bx_B}{V'} \tag{4-23}$$

with the 45° diagonal is found by solving Eqs. (4-21) and (4-23) simultaneously to give

$$x_i = y_i = \frac{Bx_B/V'}{L'/V' - 1} = x_B \tag{4-24}$$

The second half of Eq. (4-24) follows from Eq. (4-9), and thus the intersection point of the lower operating line and the 45° diagonal line is seen to occur at y and x values both equal to the bottom product composition.

4-12. Intersection of the two operating lines

In Fig. 4-13 it is seen that the intersection point for the upper and lower operating lines has a value of x equal to 0.6, which is the benzene mole fraction in the feed stream for that case. It is generally true for a saturated liquid feed that the x value of the intersection point will correspond to the feed composition. On the other hand, for a saturated vapor feed, the y value at the intersection point will equal the feed composition. In general, the locus of the point of intersection of the upper operating line and the lower operating line will depend upon the thermal condition of the feed, as represented by q, and the feed composition, z_F, which is the mole fraction of the more volatile component in the feed mixture. Since the intersection point, (y_i, x_i), will lie on both the upper and lower operating lines, it follows from Eqs. (4-12) and (4-10) that

$$Vy_i = Lx_i + Dx_D \tag{4-25}$$

$$V'y_i = L'x_i - Bx_B \tag{4-26}$$

Subtracting Eq. (4-26) from Eq. (4-25) yields

$$(V - V')y_i = (L - L')x_i + Dx_D + Bx_B \tag{4-27}$$

Substituting Eqs. (4-8), (4-13), and (4-14) into Eq. (4-27) yields

$$(1 - q)Fy_i = -qFx_i + Fz_F \tag{4-28}$$

Dividing Eq. (4-28) through by the feed rate yields the equation for the "q line,"

$$(1 - q)y_i = -qx_i + z_F \tag{4-29}$$

which describes the locus of the point of intersection of the upper and lower operating lines. It is seen from Eq. (4-29) that the "q line" is a straight line with a slope equal to $-q/(1 - q)$. The intersection of the q line with the 45° diagonal line always occurs at values of x and y equal to z_F, the feed composition, as can be demonstrated by solving Eqs. (4-21) and (4-29) simultaneously.

Figure 4-14 presents a graphical representation of Equation (4-29)

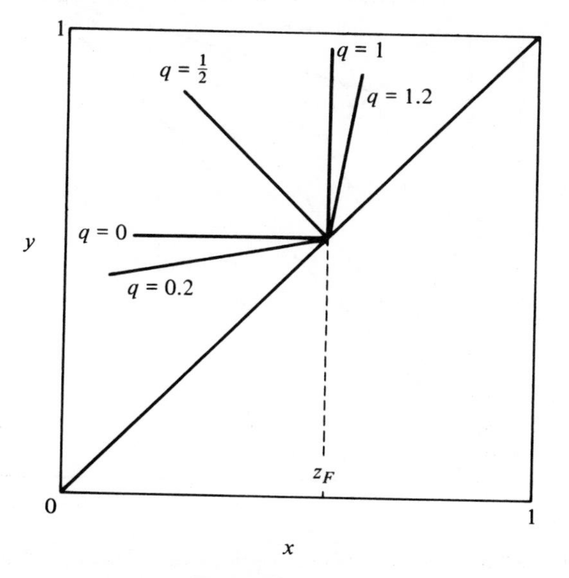

Fig. 4-14. q lines.

expressed as a plot of y_i versus x_i on the y–x diagram for several values of q. The q line always intersects the 45° diagonal line at x and y values equal to z_F. The slope of the q line is zero for a saturated vapor feed and infinity for a saturated liquid feed. For a feed that is an equimolal mixture of liquid and vapor, $q = \frac{1}{2}$, and the slope of the q line is -1. For a subcooled liquid with a q value of 1.2, the slope of the q line is 6.0. For a superheated vapor with a value of q equal to -0.2, the slope of the q line is $\frac{1}{6}$. For a given feed stream, z_F and q are fixed, and the intersection point of the upper and lower operating lines will always lie on the q line corresponding to those values of z_F and q.

4-13. Minimum reflux ratio

The effect of the reflux ratio upon the positions of the operating lines is now readily visualized with the aid of the generalities which have been derived. Figure 4-15 shows how the positions of the upper and lower operating lines vary with the reflux ratio for fixed feed and product compositions and a fixed value of q. The upper operating line always intersects the 45° diagonal at the point corresponding to the distillate product composition, and the lower operating line always intersects the 45° diagonal at the point

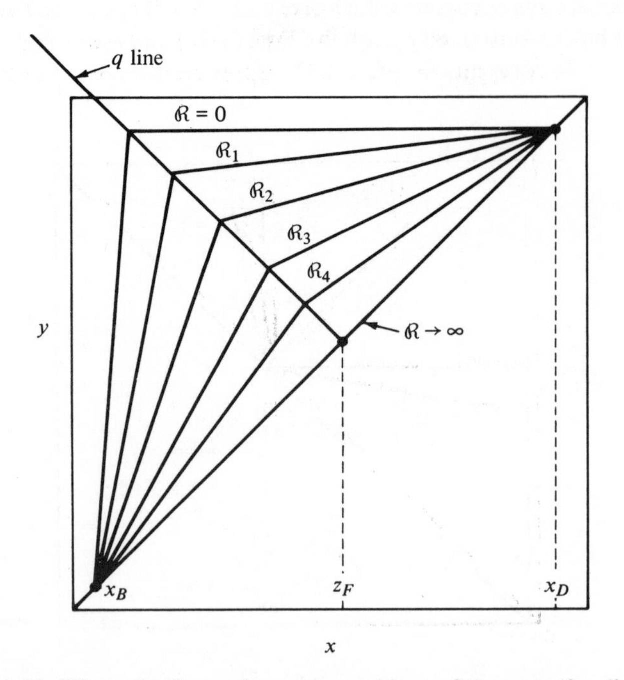

Fig. 4-15. Effect of reflux ratio on the positions of the operating lines; $\mathcal{R}_4 > \mathcal{R}_3 > \mathcal{R}_2 > \mathcal{R}_1$.

corresponding to the bottom product composition. The upper and lower operating lines intersect each other along the q line, which is drawn with a slope of -1 in Fig. 4-15, corresponding to an equimolal mixture of liquid and vapor in the feed stream, that is, $q = \frac{1}{2}$. For a reflux ratio of zero, the upper operating line is horizontal, and the lower operating line is quite steep. As the reflux ratio is increased, the slope of the upper operating line increases and approaches unity from beneath, while the slope of the lower operating line decreases and approaches unity from above. As the reflux ratio is increased toward infinity, the upper and lower operating lines approach the 45° diagonal line.

For the case shown in Fig. 4-15, when the reflux ratio is zero, the upper operating line is horizontal. At this point the slope of the lower operating line is quite high but still positive. This point corresponds to a situation in which no liquid reflux is being added to the top of the column and, therefore, all the overhead vapor is being condensed and removed as distillate product. But it is clear from Eq. (4-19) that vapor is being produced in the reboiler, because the slope of the lower operating line is positive and not equal to infinity. This corresponds to a case in which the vapor fraction of the feed stream is less than the distillate product rate; therefore, some of the liquid flowing down the column in the stripping section (which all comes from the liquid fraction of the feed) is reboiled in order that the vapor rate flowing up the column in the enriching section can just equal the distillate product rate. This corresponds to the lowest possible reflux rate which could be used in this case. On the other hand, with a lower value of q than that shown in Fig. 4-15, the q line would be more nearly horizontal, and a point would be reached at which a horizontal upper operating line would intersect the q line at a value of x smaller than x_B, with the result that the lower operating line would have a negative slope. This would correspond to an impossible situation because it would require a negative vapor rate in the bottom of the column, according to Eq. (4-19). In this case, the liquid fraction of the feed stream would be found to be less than the required bottom product rate, and thus it would be absolutely necessary that some reflux liquid be returned to the top of the column in order that the liquid flowing down the column in the stripping section should at least equal the required bottom product rate. In this case, the lowest possible liquid reflux rate would correspond to a vertical lower operating line and an upper operating line with a small positive slope. Thus the uppermost operating lines in the family shown in Fig. 4-15 will correspond to a horizontal upper operating line and a lower operating line with a large positive slope, such as is shown in Fig. 4-15, for large values of q. With small values of q the uppermost operating lines will correspond to a vertical lower operating line and an upper operating line with a small positive slope. Neither operating line can ever have a negative slope.

The uppermost lines in the family of operating lines shown in Fig. 4-15 will not correspond to the minimum reflux ratio which can be employed to achieve a given separation unless the system to be separated has an extremely high relative volatility. More generally, the minimum reflux ratio that can be employed is determined by the "pinching in" of the operating lines on the equilibrium curve. Figure 4-16 shows a family of operating lines for different reflux ratios and also an equilibrium curve for the system to be separated. As in Fig. 4-15, the operating lines move away from the 45° diagonal as the reflux ratio is decreased. But the number of steps between the operating lines and the equilibrium curve required to advance from the bottom product composition up to the distillate product composition obviously increases as the reflux ratio is decreased and the operating lines approach the equilibrium curve. Finally, as the intersection point between the upper and lower operating lines actually approaches and touches the equilibrium curve, the number of theoretical plates required to achieve the separation approaches infinity. The reflux ratio at that point, $\mathscr{R}_1$ in Fig. 4-16, is the minimum reflux ratio that could be used, even with an infinite number of theoretical plates, to achieve the desired separation. Of course, no real distillation column could employ a reflux ratio quite that low. But the reflux ratio that a real column could employ could be made to approach this minimum reflux ratio as

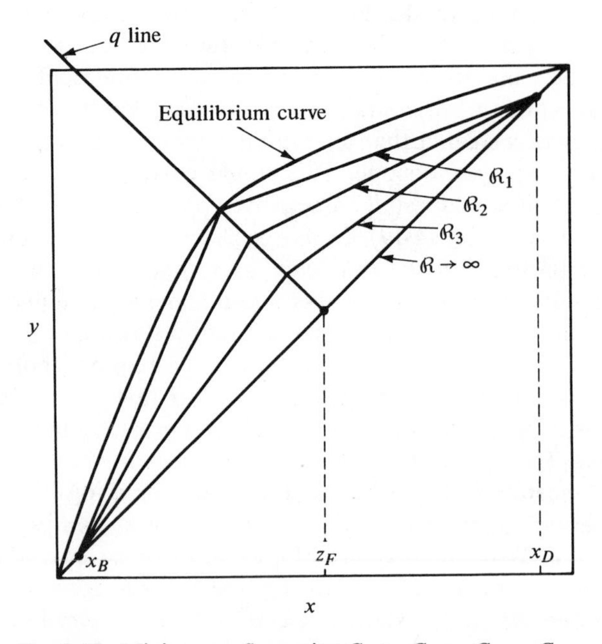

Fig. 4-16. Minimum reflux ratio; $\mathscr{R}_3 > \mathscr{R}_2 > \mathscr{R}_1 = \mathscr{R}_{\min}$.

closely as desired by employing more and more plates in the distillation column.

The minimum reflux ratio is thus determined by a "pinch" between the equilibrium curve and the operating lines, which results in the number of theoretical plates approaching infinity as the reflux ratio approaches this minimum value. When the equilibrium curve has no inflection points, this pinch will occur at the intersection of the q line and the equilibrium curve. For fixed values of z_F and q and for a given equilibrium curve, the intersection point between the q line and the equilibrium curve can be determined simply on the McCabe–Thiele diagram. From the coordinates of the intersection point together with the distillate composition, the slope of the upper operating line can be computed, and the minimum reflux ratio can then be determined from Eq. (4-18).

When the equilibrium curve has a point of inflection, the pinch between the operating line and the equilibrium curve may occur at a point of tangency rather than at the intersection of the q line and the equilibrium curve. Figure 4-17 illustrates a *tangent pinch* between the upper operating line and the equilibrium curve. It is obvious that, as the reflux ratio is decreased toward the value corresponding to the operating lines shown in Fig. 4-17, the number of steps required to get past the pinch point approaches infinity. Therefore, the operating lines shown correspond to the minimum reflux ratio that could

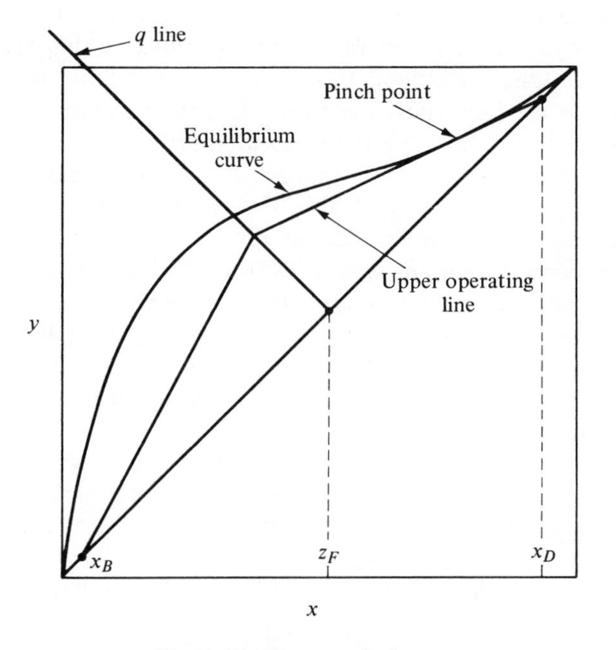

Fig. 4-17. Tangent pinch.

be used. In this case, as the reflux ratio is decreased, the pinch occurs at the point of tangency rather than at the intersection with the q line. A tangent pinch could also occur between the lower operating line and the equilibrium curve if the shape of the equilibrium curve were such that it had a positive second derivative at its lower end.

4-14. Total reflux

Referring to Fig. 4-16, the number of theoretical plates required to achieve the specified separation approaches infinity as the reflux ratio approaches its minimum value, $\mathscr{R}_1$. As the reflux ratio is increased from this value, the number of theoretical plates required decreases continuously, finally approaching an asymptote as the reflux ratio approaches infinity. This minimum number of theoretical plates that could be employed at a reflux ratio approaching infinity is an important bound on the distillation-column design. For fixed feed and product rates, a reflux ratio approaching infinity implies liquid and vapor rates within the column that are approaching infinity, and no real column operation could correspond to this limit. But, of course, the limit could be approached as closely as one desired. Furthermore, a real column is sometimes operated at *total reflux* in which feed introduction and product removal are stopped, and all the overhead vapor is condensed and returned as reflux and all the liquid from the bottom plate is reboiled and returned as vapor to the column. The concentration profiles through a column operated at total reflux in this manner are the same as those that correspond to the limit of an infinite boil-up rate and finite feed and product rates; therefore, the limiting case of infinite boil-up and reflux are commonly referred to as total reflux.

As the reflux ratio approaches infinity, the upper and lower operating lines approach the 45° diagonal line. Therefore, the number of theoretical plates required at total reflux can be obtained by steps between the 45° diagonal line and the equilibrium curve from the bottom composition up to the distillate product composition. The number of steps thus obtained corresponds to the minimum number of theoretical plates plus one, assuming that the reboiler accomplishes one theoretical stage of separation because the vapor from the reboiler is assumed to be in equilibrium with the bottom product composition.

For the special case in which the components to be separated have a constant relative volatility, the minimum number of theoretical plates required at total reflux can be determined analytically, as shown by Fenske [1]. Referring to Fig. 4-12, assuming that the vapor leaving the reboiler is in equilibrium with the bottom product liquid, Eq. (4-5) applies and

$$\frac{y_0}{1 - y_0} = \alpha \frac{x_B}{1 - x_B} \tag{4-30}$$

At total reflux both the upper and lower operating lines are described by Eq. (4-21), and thus

$$x_1 = y_0 \tag{4-31}$$

Combining Eqs. (4-30) and (4-31) yields

$$\frac{x_1}{1 - x_1} = \alpha \frac{x_B}{1 - x_B} \tag{4-32}$$

The equilibrium relationship for the bottom plate is

$$\frac{y_1}{1 - y_1} = \alpha \frac{x_1}{1 - x_1} \tag{4-33}$$

Substitution of Eq. (4-32) into Eq. (4-33) yields

$$\frac{y_1}{1 - y_1} = \alpha^2 \frac{x_B}{1 - x_B} \tag{4-34}$$

The material balance again gives

$$x_2 = y_1 \tag{4-35}$$

Therefore, Eq. (4-34) can be written

$$\frac{x_2}{1 - x_2} = \alpha^2 \frac{x_B}{1 - x_B} \tag{4-36}$$

Combining Eq. (4-36) with the equilibrium relationship for the second plate,

$$\frac{y_2}{1 - y_2} = \alpha \frac{x_2}{1 - x_2} \tag{4-37}$$

and the material balance,

$$x_3 = y_2 \tag{4-38}$$

yields

$$\frac{x_3}{1 - x_3} = \alpha^3 \frac{x_B}{1 - x_B} \tag{4-39}$$

Continuing this procedure up the column to the top plate gives

$$\frac{x_N}{1 - x_N} = \alpha^N \frac{x_B}{1 - x_B} \tag{4-40}$$

and, for the vapor leaving the top plate,

$$\frac{y_N}{1 - y_N} = \frac{x_D}{1 - x_D} = \alpha^{N+1} \frac{x_B}{1 - x_B} \tag{4-41}$$

Equation (4-41) can be written as

$$\mathcal{S} \equiv \frac{x_D/(1 - x_D)}{x_B/(1 - x_B)} \equiv \frac{Dx_D/Bx_B}{D(1 - x_D)/B(1 - x_B)} = \alpha^{N+1} \tag{4-42}$$

The left-hand side of Eq. (4-42) can be seen to represent the mole ratio of the more volatile component to the less volatile component in the distillate product divided by this ratio in the bottom product. It is also equal to the distillate-to-bottom flow-rate ratio for the more volatile component divided by this same ratio for the less volatile component. Thus the left-hand side of Eq. (4-42) represents the *separating power* of the distillation system, and it is given the symbol $\mathscr{S}$. It is analogous to the separating power defined by Eq. (3-60) for the liquid–liquid extraction cascade. Furthermore, writing Eq. (3-81) for each of two components being separated by a liquid–liquid extraction cascade and dividing the two equations would yield an equation analogous to Eq. (4-42). Taking the logarithm of both sides of Eq. (4-42) yields the Fenske equation for the minimum number of theoretical plates at total reflux required for a given separation:

$$N_{\min} + 1 = \frac{\ln \mathscr{S}}{\ln \alpha} \equiv \frac{\ln\left[\dfrac{x_D/(1 - x_D)}{x_B/(1 - x_B)}\right]}{\ln \alpha} \tag{4-43}$$

This applies to a column with an equilibrium partial reboiler and a total condenser. When a partial condenser is used, another stage of separation is achieved, as discussed later, and the left-hand side of Eq. (4-43) should be $N + 2$. It was assumed in the derivation that the vapor–liquid equilibrium curve could be described by Eq. (4-5) with a constant value of α. If the relative volatility, α, varies somewhat with composition, the geometric mean of the α values at the distillate and bottom product compositions can be used in Eq. (4-43) as a useful approximation.

EXAMPLE PROBLEM 4-3 Referring to Example Problem 4-2, what is the minimum reflux ratio that could be used to accomplish the desired separation? What is the corresponding reboiler vapor rate per 100 moles of feed? Repeat these calculations for the case of a saturated vapor feed instead of a saturated liquid feed.

Solution: For the saturated liquid feed, the q line is a vertical line at $x = 0.6$. The intersection of this q line with the equilibrium curve can be read from Fig. 4-13; the value of y at the intersection point is 0.79. The upper operating line passes through this point and also intersects the 45° diagonal line at the distillate composition of 90 mole per cent benzene. Therefore, the slope of this upper operating line is equal to

$$\frac{0.9 - 0.79}{0.9 - 0.6} = 0.3667$$

This slope is equal to the liquid-to-vapor flow-rate ratio in the enriching section of the column and is related to the reflux ratio by Eq. (4-18):

$$\frac{\mathscr{R}}{\mathscr{R} + 1} = \frac{L}{V} = 0.3667$$

Solving for the minimum reflux ratio,

$$\mathscr{R}_{\min} = \frac{0.3667}{1 - 0.3667} = 0.579$$

From the solution to Example Problem 4-1, the distillate rate on the basis of 100 moles of feed is equal to 64.7. Therefore,

$$L = \mathscr{R}D = (0.579)(64.7) = 37.5$$

From Eq. (4-11), the vapor rate is calculated to be

$$V = 37.5 + 64.7 = 102.2$$

For the liquid feed, $q = 1$, and

$$V' = V = 102.2$$

Thus the minimum reflux ratio is 0.579, corresponding to 102.2 moles of vapor reboiled per 100 moles of feed.

If the feed were a saturated vapor rather than a saturated liquid, the value of q would be zero and the q line would be a horizontal line at $y = 0.6$. From Fig. 4-13 it is seen that this line intersects the equilibrium curve at $x = 0.38$. The upper operating line passes through this intersection point and also intersects the 45° diagonal line at $x = 0.9$. Therefore, the slope of the upper operating line is

$$\frac{L}{V} = \frac{0.9 - 0.6}{0.9 - 0.38} = 0.577$$

From Eq. (4-18) it follows that the minimum reflux ratio is

$$\mathscr{R}_{\min} = \frac{0.577}{1 - 0.577} = 1.363$$

The reflux rate is

$$L = (1.363)(64.7) = 88.2$$

and the vapor rate in the enriching section is given by Eq. (4-11).

$$V = 88.2 + 64.7 = 152.9$$

Using Eq. (4-14) with $q = 0$, the vapor rate in the stripping section is

$$V' = 152.9 - 100 = 52.9$$

Therefore, with the saturated vapor feed the minimum reflux ratio is 1.363, corresponding to a reboiler vapor rate of 52.9 moles per 100 moles of feed. It should be noted that with the vapor feed, the minimum reflux ratio is greater than that required with a liquid feed, but the reboiler vapor rate requirement is lower with the vapor feed.

EXAMPLE PROBLEM 4-4 Referring to Example Problem 4-2, what is the minimum number of theoretical plates that could be used to accomplish the desired separation?

Solution: The minimum number of theoretical plates corresponds to the limit of an infinite reflux ratio, or total reflux. It can be obtained graphically on a McCabe–Thiele diagram in which the 45° diagonal line represents both the upper and the lower operating lines. Starting with $x_B = 0.05$ and stepping up to the equilibrium curve, over to the 45° diagonal line, up to the equilibrium curve, back over to the 45° diagonal line, and continuing up the column, the steps are determined as shown in Fig. 4-18. Assuming that the vapor leaving the partial reboiler is in equilibrium with

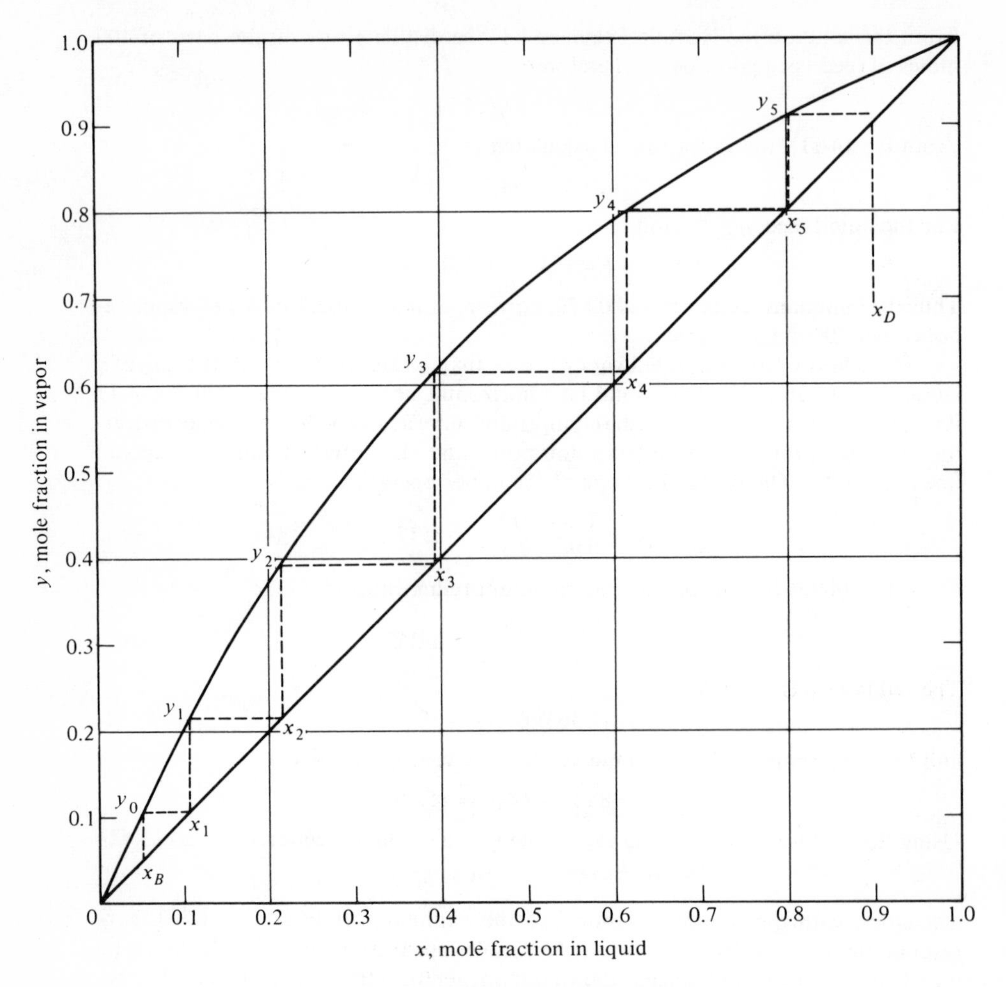

Fig. 4-18. Solution to Example Problem 4-4.

the bottom product composition, it is seen that the vapor leaving the fifth theoretical plate is slightly richer than 90 mole per cent benzene, the required distillate product composition. Therefore, just under six theoretical stages are required, or the reboiler plus just under five theoretical plates.

Since the relative volatility for the benzene–toluene system is nearly constant, the minimum number of theoretical plates may also be estimated from Eq. (4-43). Reading the equilibrium curve in Fig. 4-13 at $x = 0.5$, the equilibrium value of y is 0.713. The relative volatility follows from Eq. (4-5):

$$\alpha = \frac{0.713/0.5}{0.287/0.5} = 2.48$$

At $x = 0.1$, the equilibrium value of y is 0.203. Again using Eq. (4-5),

$$\alpha = \frac{0.203/0.1}{0.797/0.9} = 2.29$$

At $x = 0.9$, $y = 0.958$, and

$$\alpha = \frac{0.958/0.9}{0.042/0.1} = 2.53$$

The geometric mean of the relative volatilities at $x = 0.1$ and at $x = 0.9$ is

$$\bar{\alpha} = \sqrt{(2.29)(2.53)} = 2.41$$

which is reasonably close to the value of $\alpha = 2.48$ at $x = 0.5$. The minimum number of theoretical plates is calculated from Eq. (4-43). Using a value of $\alpha = 2.41$ gives

$$N_{\min} = \frac{\ln\left[\frac{0.9/0.1}{0.05/0.95}\right]}{\ln 2.41} - 1 = 4.85$$

This value is in good agreement with the one determined graphically; an equilibrium reboiler plus just under five theoretical plates are required at total reflux.

EXAMPLE PROBLEM 4-5 A mixture of acetone and water contains 55 mole per cent acetone and 45 mole per cent water. It is desired to distill this feed stream into a distillate product that will contain 98 per cent of the acetone fed to the distillation column at a purity of at least 96 mole per cent acetone. The feed stream is a mixture of 60 per cent liquid and 40 per cent vapor, and the reflux ratio to be employed is 1.5. How many theoretical plates are required to effect this separation? To which plate should the feed stream be added to minimize the number of plates required? The distillation column will operate at a pressure of 1 atm and will contain a partial reboiler and a total condenser.

Solution: On the basis of 100 moles of feed, 55 moles of acetone and 45 moles of water are fed to the column. The moles of acetone in the distillate product are to be 98 per cent of the 55 moles fed, or 53.9 moles in the distillate product. The remainder will appear in the bottom product stream. Since the distillate product is to be at least 96 mole per cent acetone, the moles of water in the distillate product will be

$$\tfrac{4}{96}(53.9) = 2.245$$

The remainder of the 45 moles of water will appear in the bottom product stream. On this basis, the overall material balance is readily worked out as shown in Table 4-1. The distillate product is 96 mole per cent acetone as specified, and the acetone

Table 4-1

	Molal Flow Rate (moles/unit time) in		
Component	*Feed*	*Distillate*	*Bottom Product*
Acetone	55	53.9	1.1
Water	45	2.245	42.755
Total	100	56.145	43.855

mole fraction in the bottom product stream is

$$x_B = \frac{1.1}{43.855} = 0.0251$$

For a reflux ratio of 1.5, the slope of the upper operating line is computed according to Eq. (4-18) as

$$\frac{L}{V} = \frac{1.5}{2.5} = 0.6$$

Therefore, the upper operating line is located by drawing a line with a slope of 0.6 that intersects the 45° diagonal line at the distillate product composition, $x_D = 0.96$, as shown in Fig. 4-19. For this feed stream, which is 40 per cent vaporized, $q = 0.6$. The q line is represented by Eq. (4-29); its slope is given by

$$\frac{-q}{1-q} = \frac{-0.6}{0.4} = -1.5$$

Therefore, the q line is located by drawing a line with a slope of -1.5 that intersects the 45° diagonal line at the feed composition, 55 mole per cent acetone, as shown in Fig. 4-19. The q line and the upper operating line are found to intersect at the point ($y = 0.665$, $x = 0.47$). The lower operating line must pass through this point and must intersect the 45° diagonal line at $x_B = 0.0251$; thus it is readily located, as shown in Fig. 4-19.

Although not necessary in this case, it is useful to calculate the internal flow rates of liquid and vapor within the column. The liquid reflux rate is given by the product of the reflux ratio and the distillate rate:

$$L = (1.5)(56.145) = 84.22$$

The vapor rate in the enriching section is given by Eq. (4-11) as

$$V = 56.145 + 84.22 = 140.4$$

Equations (4-13) and (4-14) then yield the liquid and vapor flow rates in the stripping section:

$$L' = 84.22 + 60 = 144.22$$

$$V' = 140.4 - 40 = 100.4$$

It is seen that Eq. (4-9) is satisfied. The slope of the lower operating line is

$$\frac{L'}{V'} = \frac{144.22}{100.4} = 1.435$$

This slope is in good agreement with the slope of the line constructed graphically on the McCabe–Thiele diagram.

Using these operating lines and the given equilibrium curve for the acetone–water system at a pressure of 1 atm, the steps are constructed as shown in Fig. 4-19. It is assumed that the vapor leaving the reboiler is in equilibrium with the bottom product composition. The point (y_0, x_1) lies on the lower operating line, and the point (y_1, x_2) lies on the upper operating line. Therefore, the feed stream is to be fed to the first theoretical plate. According to the construction in Fig. 4-19, y_7 is approximately equal to 0.96; therefore, an equilibrium reboiler plus seven theoretical plates are required.

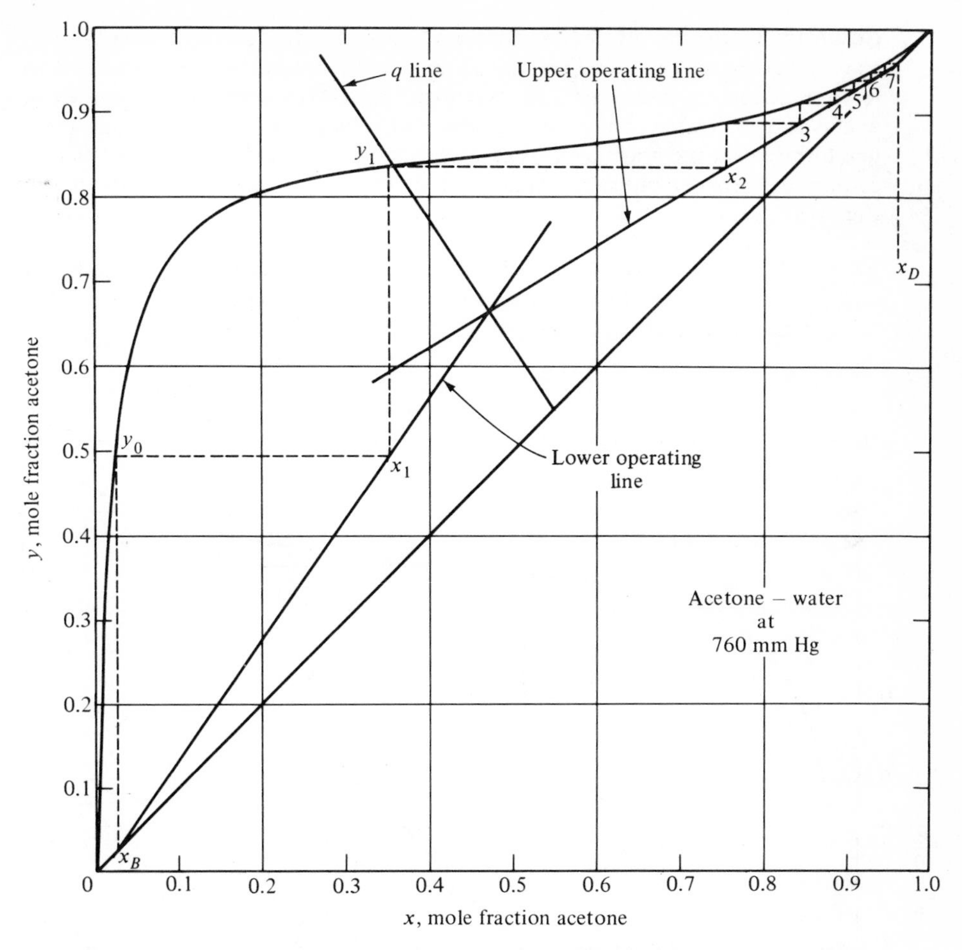

Fig. 4-19. Solution to Example Problem 4-5.

The graphical construction shown is not an accurate plate count, especially in the upper right-hand corner of the diagram, and somewhat more accuracy could be obtained by expanding the diagram. If very high purities are desired, a greatly expanded diagram is needed and often logarithmic plotting of the equilibrium curve and the operating line is employed. Alternatively, using a mathematical equation fit to the equilibrium curve, the computation can be carried out arithmetically on a digital computer. But Fig. 4-19 is adequate for illustrating the solution in this case.

EXAMPLE PROBLEM 4-6 Referring to Example Problem 4-5, what is the minimum number of theoretical plates that could be used to accomplish the desired separation?

Solution: Inspection of the equilibrium curve for the acetone–water system reveals that the relative volatility is much greater at low acetone mole fractions than it is at high acetone mole fractions. Therefore, Eq. (4-43) is probably not very useful in this case. A plate count on a McCabe–Thiele diagram, using the 45° diagonal line to represent the upper and lower operating lines at total reflux, shows that an equilibrium reboiler plus approximately four theoretical plates are required (see Fig. 4-20).

EXAMPLE PROBLEM 4-7 Referring to Example Problem 4-5, what is the minimum reflux ratio that can be used to accomplish the separation? What is the corresponding reboiler vapor rate per 100 moles of feed?

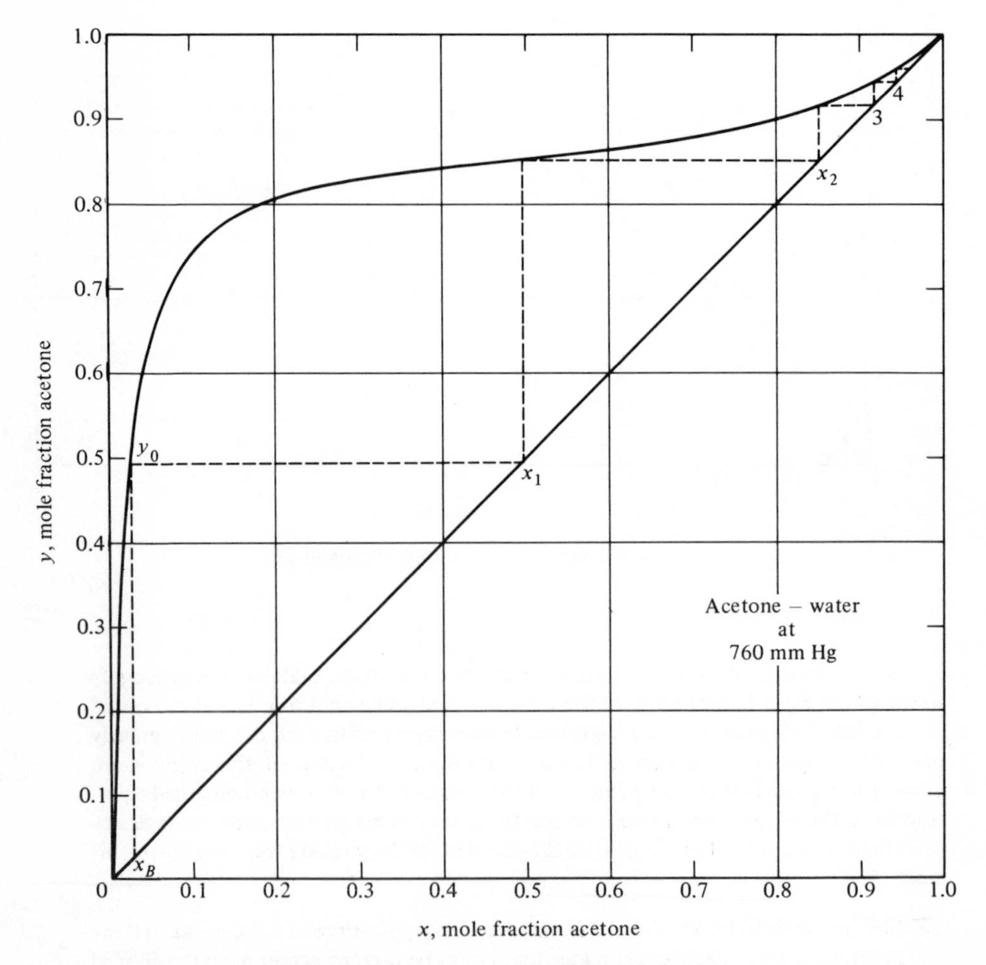

Fig. 4-20. Solution to Example Problem 4-6.

Solution: Because of the inflection point in the equilibrium curve, a tangent pinch is suspected. Therefore, the upper operating line is constructed so that it is tangent to the equilibrium curve while intersecting the 45° diagonal line at $x = 0.96$, as shown in Fig. 4-21. The q line is the same as it was in Example Problem 4-5; therefore the lower operating line is located readily by the intersection of the upper operating line with the q line, as shown in Fig. 4-21. It is obvious from the diagram that the minimum reflux ratio does indeed correspond to a tangent pinch. The upper operating line and the q line intersect at the point ($y = 0.738$, $x = 0.423$). The slope of the upper operating line is then computed as

$$\frac{L}{V} = \frac{0.96 - 0.738}{0.96 - 0.423} = 0.413$$

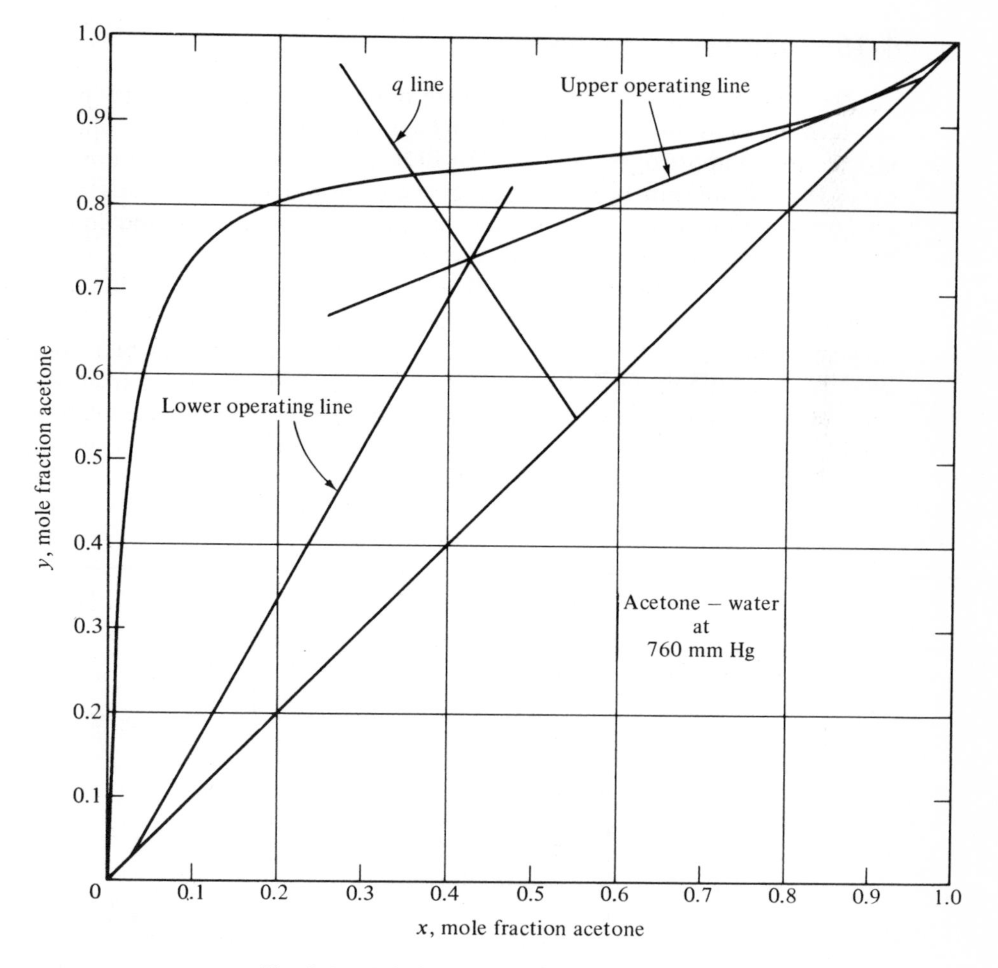

Fig. 4-21. Solution to Example Problem 4-7.

From Eq. (4-18), the reflux ratio is computed as

$$\mathscr{R}_{\min} = \frac{0.413}{1 - 0.413} = 0.704$$

which is the minimum reflux ratio that can be used. The reflux rate is

$$L = (0.704)(56.145) = 39.5$$

and the vapor rates in the enriching and stripping sections are

$$V = 39.5 + 56.145 = 95.65$$

$$V' = 95.65 - (0.4)(100) = 55.65$$

Therefore, the minimum reboiler vapor rate is 55.65 moles of vapor per 100 moles of feed.

4-15. Feed-plate location

The location of the feed plate determines the location of the point at which the steps on a McCabe–Thiele diagram shift from the lower operating line to the upper operating line. It will be instructive to consider in some detail the relationship between the method of introducing the feed stream and the position of the steps on the McCabe–Thiele diagram and indeed the assumptions implicit in this relationship.

A liquid feed stream is often introduced into a distillation column by adding it to the downpipe carrying the liquid overflow from plate $f + 1$ down to plate f, the feed plate. Thus the liquid fed to the feed plate is a mixture of the liquid feed stream and the liquid-overflow stream from the plate above. Figure 4-22 shows a sketch of this arrangement. Since the feed liquid is introduced into the downpipe, there is no contact between the

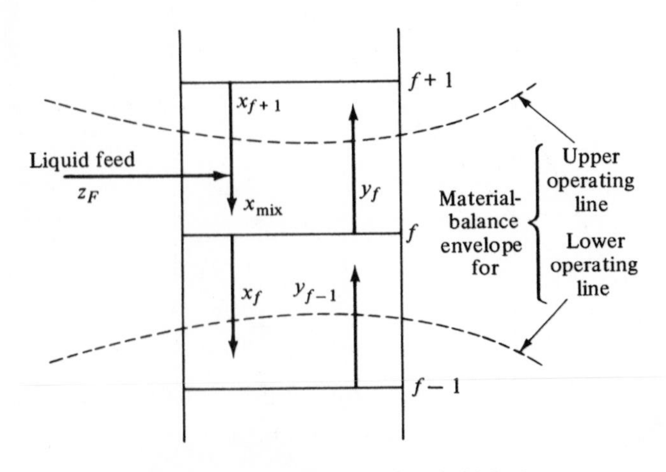

Fig. 4-22. Introduction of a liquid feed.

feed liquid and any vapor stream until the mixture of the feed liquid and the overflow from the plate above contacts the vapor stream on plate f. The vapor rising from plate f is assumed to be in equilibrium with the liquid overflowing from that plate, the standard assumption for a theoretical plate. Figure 4-22 shows by dashed lines the material-balance envelopes for the upper operating line and the lower operating line. In defining these envelopes it is important that the feed stream not be included in the material balance around the top of the column for the upper operating line, or around the bottom of the column for the lower operating line, in order that the material-balance relationships may be represented by Eqs. (4-10) and (4-12). It is seen in Fig. 4-22 that the point (y_{f-1}, x_f) should lie on the lower operating line, but the point (y_f, x_{f+1}) should lie on the upper operating line. Figure 4-23 shows the corresponding McCabe–Thiele diagram. The point (y_{f-1}, x_f) lies on the lower operating line. After stepping up to the equilibrium curve to locate y_f, the point (y_f, x_{f+1}) is located on the upper operating line. Thus this step on the McCabe–Thiele diagram steps from the lower operating line to the upper operating line. The value of x at the last point on the

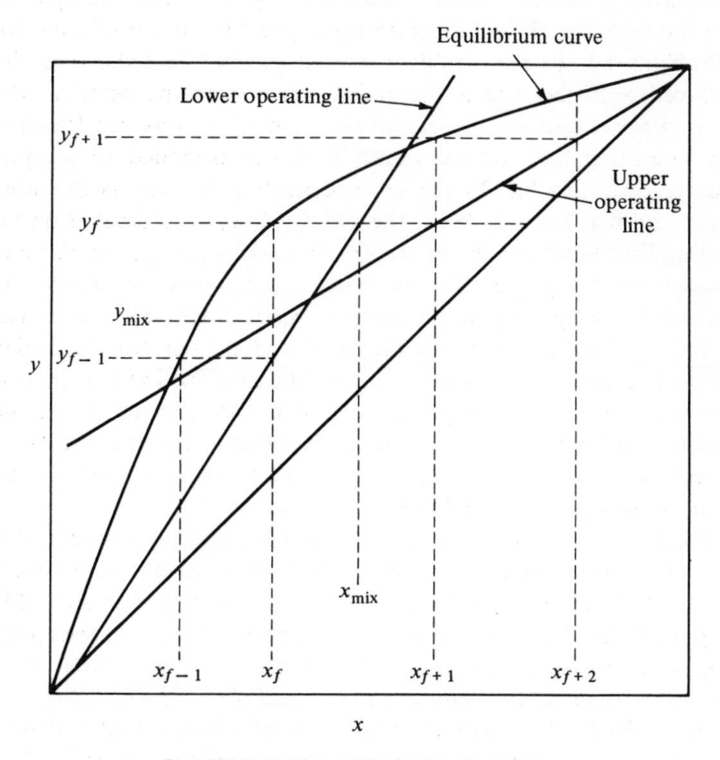

Fig. 4-23. Optimum feed-plate location.

lower operating line corresponds to x_f, the composition of the liquid flowing from the feed plate, which is defined as the plate *to which* the feed stream flows. It can be seen in Fig. 4-22 that the point (y_f, x_{mix}) will lie on the lower operating line, because a material balance including those two streams and the bottom of the column would not include the feed stream. Therefore, the composition of the mixture of the feed liquid and the overflow liquid from plate $f + 1$ can be located on the lower operating line at the value of y_f, as indicated in Fig. 4-23.

The McCabe–Thiele diagram in Fig. 4-23 illustrates the *optimum feed-plate location.* Here the feed plate was chosen such that a McCabe–Thiele step reached from the lower operating line to the upper operating line by stepping across the intersection between the two operating lines. This determines the feed-plate location such that the total number of theoretical plates is minimized. Although this is a logical choice for the design of a distillation column, it should not be inferred that this choice is necessary. What is absolutely necessary in the construction of a McCabe–Thiele diagram is that the last point on the lower operating line should be the point (y_{f-1}, x_f) and the first point on the upper operating line should be the point (y_f, x_{f+1}), where plate f is the feed plate, as shown in Fig. 4-22. For example, Fig. 4-24 shows the McCabe–Thiele diagram corresponding to the solution to Example Problem 4-1. In this problem it was specified that the feed should be introduced onto the eighth theoretical plate from the bottom. Therefore, after y_5 was located on the equilibrium curve, x_6 was not located on the upper operating line, which would have corresponded to stepping from the lower operating line to the upper operating line across the intersection between the two lines. Instead, the point (y_5, x_6) was located on the lower operating line. Similarly, the points (y_6, x_7) and (y_7, x_8) were also located on the lower operating line. But the point (y_8, x_9) was located on the upper operating line because plate 8 was specified as the feed plate. Therefore, the McCabe–Thiele diagram in Fig. 4-24 and the numerical calculations of Example Problem 4-1 are consistent with the specification that plate 8 should be the feed plate, even though this design is not optimum in the sense that the total number of theoretical plates required is greater than the number required in the solution to Example Problem 4-2, in which plate 5 was chosen as the optimum feed-plate location.

When the feed stream is a vapor, it is commonly introduced into the vapor space above plate $f - 1$, where it mixes with the vapor rising from plate $f - 1$ and flows up to plate f, as shown in Fig. 4-25. It should be noted in Fig. 4-25 that the vapor feed is fed beneath plate f, the feed plate, while in Fig. 4-22 the liquid feed is introduced above the feed plate. But in both instances the feed stream is *flowing to* the feed plate, flowing down to the feed plate with the liquid overflow in the case of a liquid feed and flowing up to the feed plate with the vapor stream in the case of a vapor feed. In the case

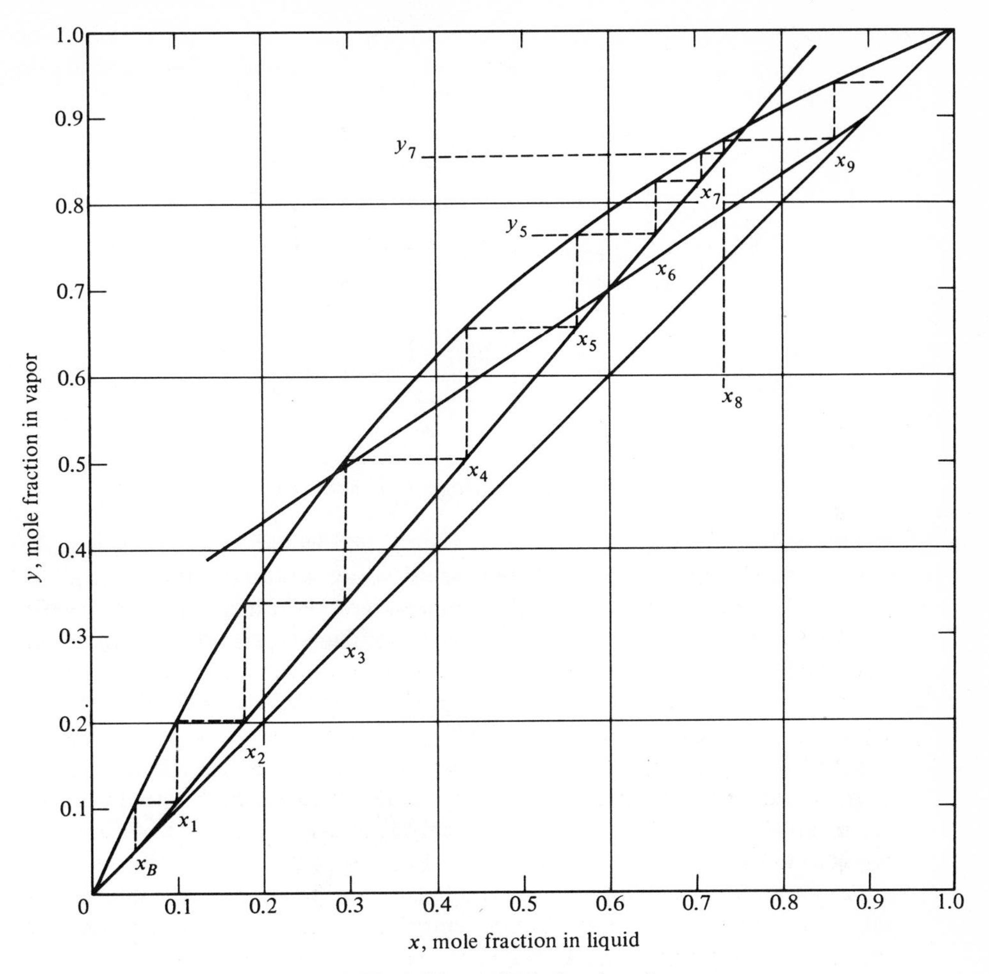

Fig. 4-24. Arbitrary feed-plate location.

of the liquid feed shown in Fig. 4-22, it was assumed that the liquid feed did not contact any vapor stream until the mixture of liquid feed and liquid overflow from the plate above contacted the vapor on the feed plate, plate *f*. In the case of the vapor feed shown in Fig. 4-25, it could also be assumed that the vapor feed does not come into intimate contact with any liquid until the mixture of the vapor feed and the vapor rising from the plate beneath contact the liquid on plate *f*. But while this assumption should be a good one for a liquid feed introduced into a downpipe, reference to Fig. 4-10 will cast considerable doubt on the assumption for the case of a vapor feed. Referring to Fig. 4-10, if a vapor feed is introduced into the vapor

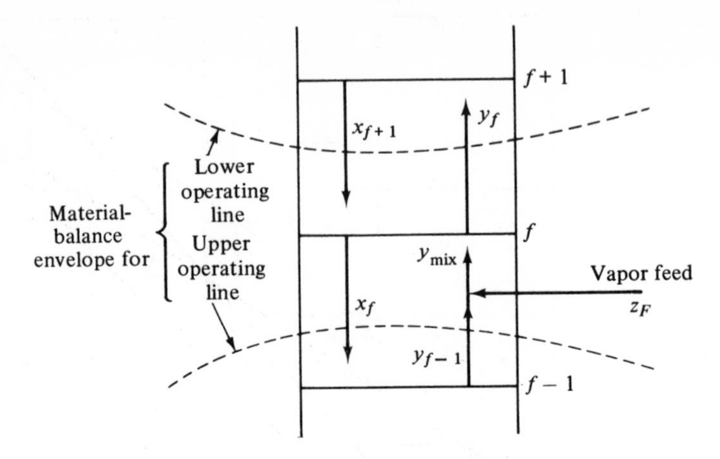

Fig. 4-25. Introduction of a vapor feed.

space above plate $f - 1$ and beneath plate f, one would expect considerable contact of the vapor feed with the liquid spray above the froth on plate $f - 1$, although the contact would probably not be as intimate as that with the liquid on plate f. Nevertheless, it is convenient to assume, according to Fig. 4-25, that the vapor feed simply mixes with the vapor leaving plate $f - 1$ and that the mixture flows up to plate f. With this simplifying assumption, Fig. 4-25 reveals that the point (y_{f-1}, x_f) lies on the lower operating line and the point (y_f, x_{f+1}) lies on the upper operating line. Therefore, the steps would be arranged on the McCabe–Thiele diagram in the same manner for a vapor feed as they would for a liquid feed, and indeed Fig. 4-23 could apply to either one, except that x_{mix} shown in Fig. 4-23 would not be relevant to the case of a vapor feed. Instead, y_{mix}, the composition of the mixture of the vapor feed and the vapor stream rising from plate $f - 1$, could be located on the upper operating line at the value x_f.

It is often tacitly assumed that a liquid feed stream flows to the feed plate, where it first contacts vapor, and a vapor feed stream flows to the feed plate, where it first contacts liquid. Therefore, the steps on the McCabe–Thiele diagram simply step from the point (y_{f-1}, x_f) on the lower operating line to the point (y_f, x_{f+1}) on the upper operating line, where plate f, the feed plate, is that plate to which the feed flows for its first contact with the other phase. This idealization may depart rather substantially from the reality of vapor–liquid contact in a real column when a vapor stream is introduced or when a liquid stream is introduced in some manner other than flowing into the downpipe. But it would be difficult to account for such effects completely, and in most cases they are probably not very important. Consequently, it is customary to design distillation columns based upon the simplifications indicated.

When the feed stream consists of a mixture of liquid and vapor, one method of introduction of the feed stream into the distillation column would be to first separate the vapor and liquid fractions of the feed stream and then to feed the liquid fraction to plate f from above and the vapor fraction to plate f from beneath, as indicated in Fig. 4-26. With this method of introduction, and with the simplifying assumptions discussed in the preceding paragraphs, it is clear from Fig. 4-26 that once again the point (y_{f-1}, x_f) lies on the lower operating line while the point (y_f, x_{f+1}) lies on the upper operating line, and as usual the point (y_f, x_f) lies on the equilibrium curve. Therefore, the location of the steps on the McCabe–Thiele diagram is the same as it would be for a feed stream that was all liquid or all vapor. Therefore, with this method of introduction of a partially vaporized feed and with the simplifications indicated, the location of the steps on the McCabe–Thiele diagram is the same for a feed stream that is all liquid, all vapor, or a mixture of liquid and vapor.

Needless to say, industrial practice does not always correspond to separating a partially vaporized feed into a liquid fraction and a vapor fraction and to introducing the liquid fraction onto the feed plate from above and introducing the vapor fraction to the feed plate from beneath. If the mixture of liquid and vapor is simply dumped into the column between two plates, the liquid fraction will flow down to the plate beneath, and the vapor fraction will flow up to the plate above. The construction of the McCabe–Thiele diagram would be modified if one wanted to model this situation as if the feed stream were separated into a liquid fraction and a vapor fraction, and the liquid fraction were fed to one plate while the vapor fraction were fed to the plate above. But the difference between considering the feed to have been fed to two plates and considering the feed to have been fed to a single plate will generally be small, and it is customary to analyze the introduction of a partially vaporized feed as if the liquid and vapor fractions were fed to the same plate.

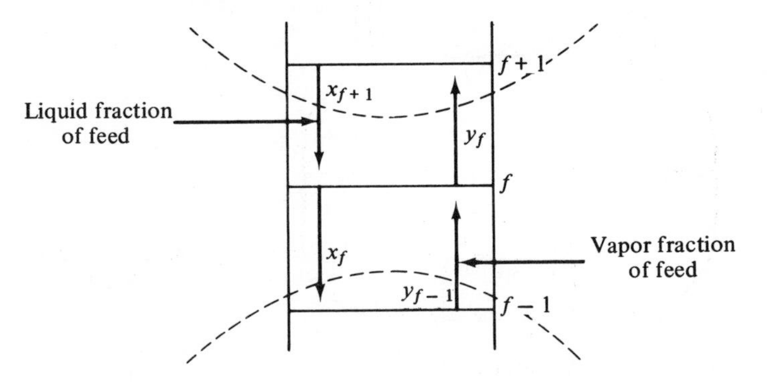

Fig. 4-26. Introduction of a partially vaporized feed.

4-16. Partial and total reboilers

Up to this point it has simply been assumed that the composition of the vapor leaving the reboiler, y_0, is in equilibrium with the liquid bottom product composition, x_B. This is a commonly employed assumption based on the concept of a partial reboiler with a well-mixed liquid holdup. Referring to Fig. 4-27, the liquid from the bottom plate in the distillation column, of composition x_1, flows into the reboiler, and the vapor from the reboiler, of composition y_0, flows back to the distillation column. The reboiler contains a quantity of liquid assumed to be well mixed, owing to the agitation caused by the boiling. The liquid composition within the reboiler is therefore assumed to be approximately uniform and equal to the composition of the liquid withdrawn as bottom product, even though the liquid flowing to the reboiler from the bottom plate in the distillation column is of a different composition. In the idealization of a completely mixed liquid holdup, drops or slugs of liquid of composition x_1 which enter the reboiler from the distillation column are assumed to be instantaneously mixed with the liquid held up in the reboiler, and thus the liquid holdup in the reboiler is visualized as being of uniform composition. Since the bottom product liquid is withdrawn from this well-mixed pool, the bottom product composition is equal to the composition everywhere within the liquid pool.

The vapor bubbles, most of which are produced at the surface of the heat source (for example, a steam-heated pipe), must bubble up in intimate contact with the liquid pool in the reboiler. It is usually assumed that this contact is sufficient to bring the vapor composition essentially to equilibrium with the liquid composition. This is only an approximation; the vapor

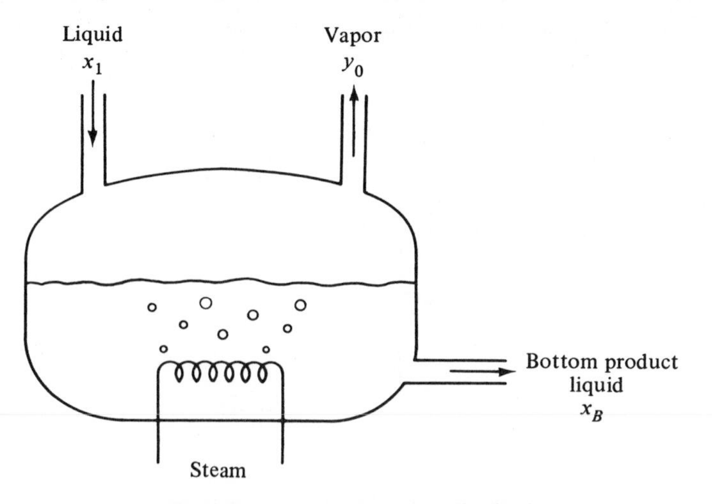

Fig. 4-27. Partial reboiler, well-mixed.

produced by boiling a liquid will not necessarily be of a composition corresponding to equilibrium with that liquid. But vapor–liquid contact in the reboiler is usually substantially better than that on the distillation plates; consequently, the approach to equilibrium in the reboiler is usually considerably closer than the approach to equilibrium on the plates. Therefore, even when a plate efficiency less than 100 per cent is employed to describe lack of equilibrium attainment on the distillation plates, it is often assumed that the vapor from the reboiler is in equilibrium with the liquid in the reboiler. On the other hand, this is an assumption that can be wrong, and for systems in which the plate efficiency is very small it should be recognized that such an assumption may not be justified. Furthermore, the assumption that the liquid holdup in the reboiler is completely mixed is surely only an approximation, but it is a widely used approximation which probably introduces little error in most cases.

It is possible to use a total reboiler instead of a partial reboiler. Figure 4-28 shows a total reboiler with a liquid bottom product stream. The liquid overflow from the bottom plate, of composition x_1, is split into a liquid bottom product stream and a liquid stream which flows to the reboiler. The rate of heat addition to the reboiler is adjusted so that the liquid holdup is constant; therefore, when in steady operation, the molal flow rate of the vapor from the reboiler is equal to the molal flow rate of the liquid to it. Furthermore, the composition of the vapor flowing from the reboiler must, in the steady state, equal the composition of the liquid flowing to the reboiler,

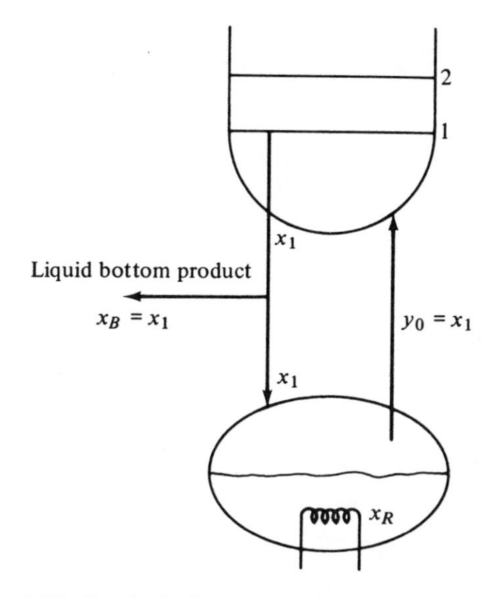

Fig. 4-28. Total reboiler with liquid bottom product.

as a simple material balance will demonstrate. In this case, the liquid holdup within the reboiler will attain some composition, x_R, at whatever value it must in order that the vapor leaving the reboiler will have a composition, y_0, equal to the composition of the liquid fed to the reboiler, x_1. Since the bottom product composition is also equal to x_1, it is seen that this type of operation of a total reboiler results in a vapor composition, y_0, which is equal to the bottom product composition rather than in equilibrium with it.

Figure 4-29 shows the arrangement for a total reboiler with a vapor bottom product. The liquid from the bottom tray is sent to the reboiler, where it is totally reboiled, and thus again the composition of the vapor leaving the reboiler will equal the composition of the liquid sent to the reboiler. The vapor stream is then split into a bottom product vapor stream and a vapor stream which is sent back to the distillation column. Therefore, in this case also the composition of the vapor fed to the bottom plate of the distillation column is equal to the bottom product composition.

The most usual method of operating a distillation column is with a liquid bottom product stream and a partial reboiler. In this case, the column design is usually based upon the assumption that the vapor sent to the bottom plate in the distillation column is in equilibrium with the liquid bottom product stream composition. But the preceding discussion reveals that if a total reboiler is employed, the composition of the vapor fed to the bottom plate in the distillation column will be equal to the composition of the bottom product, whether that product is withdrawn as a liquid or as a vapor.

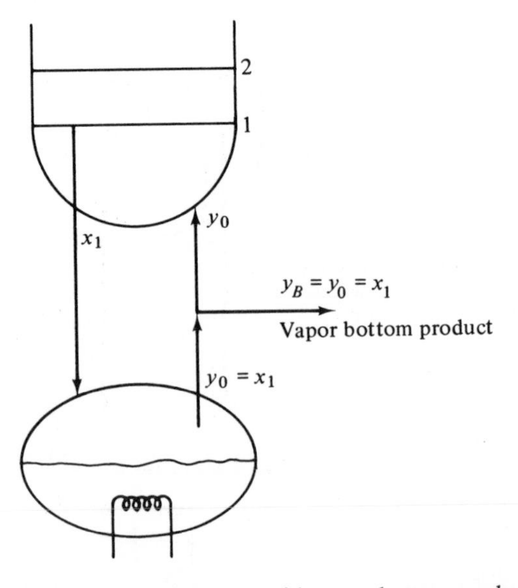

Fig. 4-29. Total reboiler with vapor bottom product.

4-17. Partial and total condensers

Up until this point it has been assumed that the composition of the vapor leaving the top plate in the distillation column is equal to the distillate product composition. This is true when a total condenser is employed. Figure 4-30 shows the arrangement of a total condenser with a liquid distillate product stream. The vapor leaving the top plate, of composition y_N, is condensed totally; therefore, the composition of the condensate must be the same as the composition of the vapor that was condensed. The liquid condensate is split into a liquid distillate product stream and a liquid reflux stream, both of which have the same composition as the vapor rising from the top plate.

Figure 4-31 shows the arrangement of a total condenser with a vapor distillate product stream. The vapor rising from the top plate is split into a vapor distillate product stream and a vapor stream which is sent to the condenser. This vapor stream is then totally condensed to form the liquid reflux stream which is returned to the top plate of the column. In this case also the distillate product and the liquid reflux stream have the same composition as the vapor rising from the top plate.

When a vapor stream is only partially condensed, the liquid condensate will generally be richer in the less volatile component while the vapor left

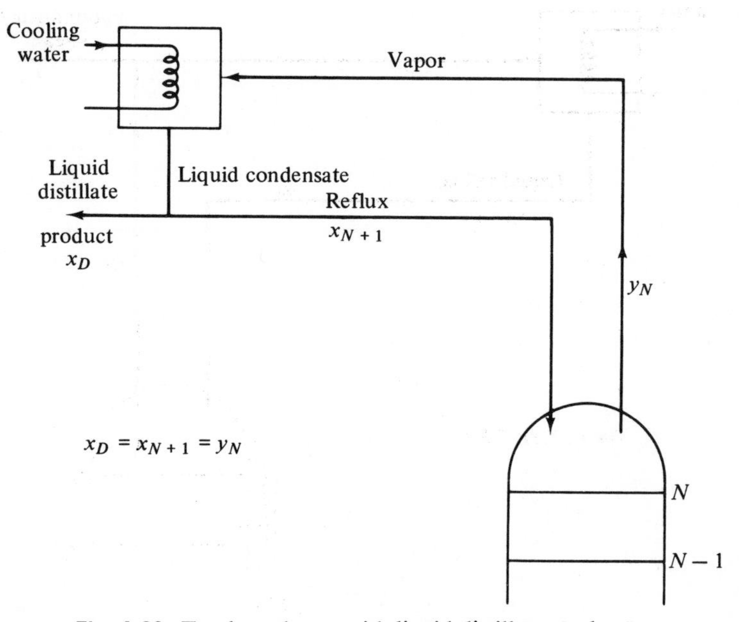

Fig. 4-30. Total condenser with liquid distillate product.

behind will generally be richer in the more volatile component; therefore, some separation of the more volatile component from the less volatile component is achieved in the partial condensation. Consequently, when it is desired to produce a vapor distillate product stream from a distillation column, a partial condenser is often employed instead of a total condenser to achieve some additional separation. Figure 4-32 shows the configuration used with a partial condenser and a vapor distillate product stream. The vapor leaving the top plate of the distillation column is sent to the condenser where it is partially condensed to produce a liquid condensate stream and a vapor distillate product stream. The liquid condensate stream is returned to the distillation column as reflux. The simplifying assumption often adopted in this case is that the vapor-distillate-product-stream composition is in equilibrium with the liquid reflux composition. This is only an approximation, the accuracy of which would depend upon the nature of the liquid and vapor flow within the condenser and the intimacy of contacting between the phases and the degree of mixing within the phases in the condenser. But it is a commonly employed assumption, and it is probably a reasonable approximation in most cases. When this simplification is adopted, y_D is assumed to be in equilibrium with x_{N+1}. Furthermore, y_N and x_{N+1} lie on the upper operating

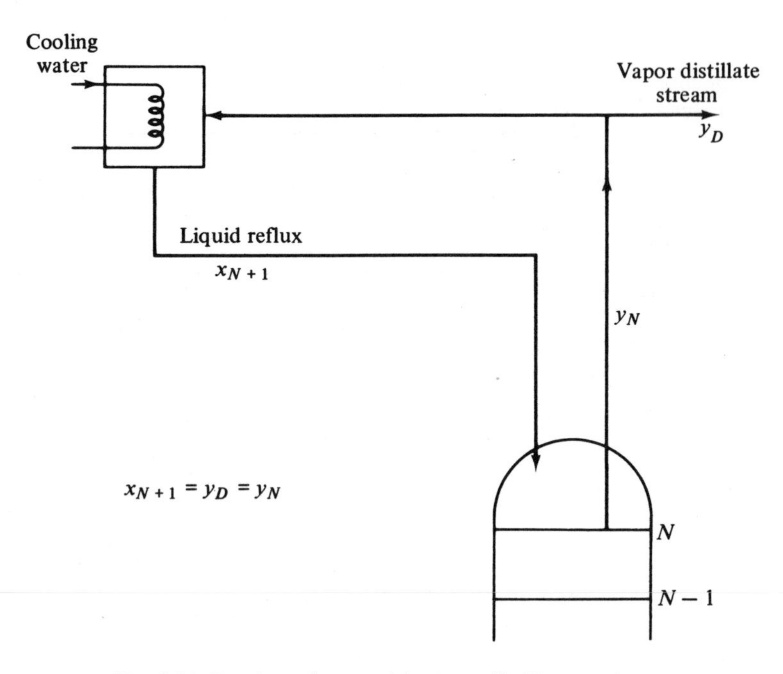

Fig. 4-31 Total condenser with vapor distillate product.

line, which at this point represents a simple material balance around the condenser.

The construction of the McCabe–Thiele diagram in this case is shown in Fig. 4-33. The upper operating line intersects the 45° diagonal line at a point corresponding to the distillate product composition, y_D in this case of a vapor distillate product. The point (y_D, x_{N+1}) is located on the equilibrium curve because of the assumption that the vapor and liquid streams leaving the partial condenser are in equilibrium. The point (y_N, x_{N+1}) lies on the upper operating line because the material-balance relationship around the condenser is the same as that employed between any plate in the enriching section and the distillate product stream. The point (y_N, x_N) is located on the equilibrium curve according to the usual assumption that the vapor and liquid streams leaving a theoretical plate are in equilibrium. The point (y_{N-1}, x_N) is located on the upper operating line as usual. It should be noted in Fig. 4-33 that, in contrast to the assumption that y_N is equal to the distillate product composition which is employed for a total condenser, an *additional step* on the McCabe–Thiele diagram is employed to represent the

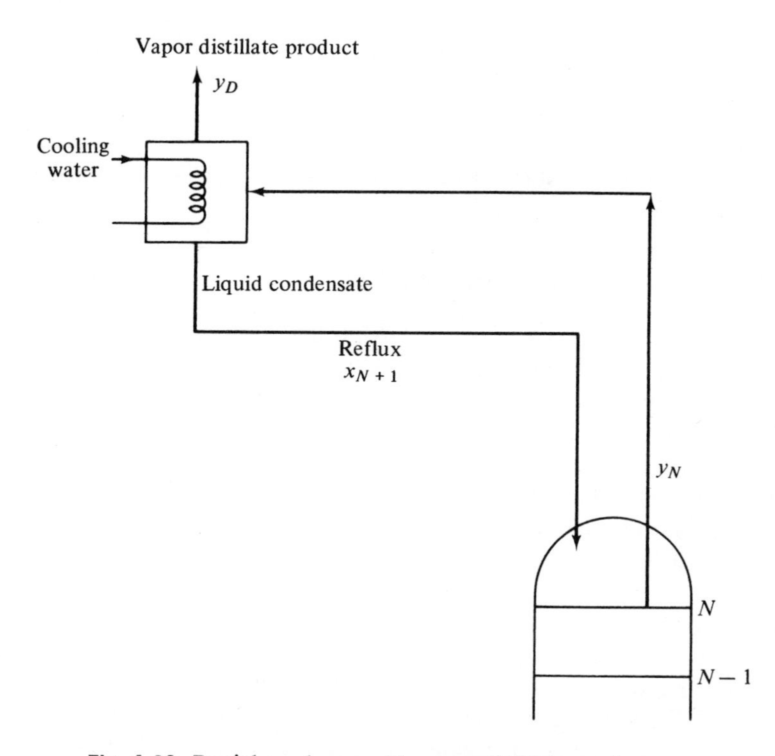

Fig. 4-32. Partial condenser with vapor distillate product.

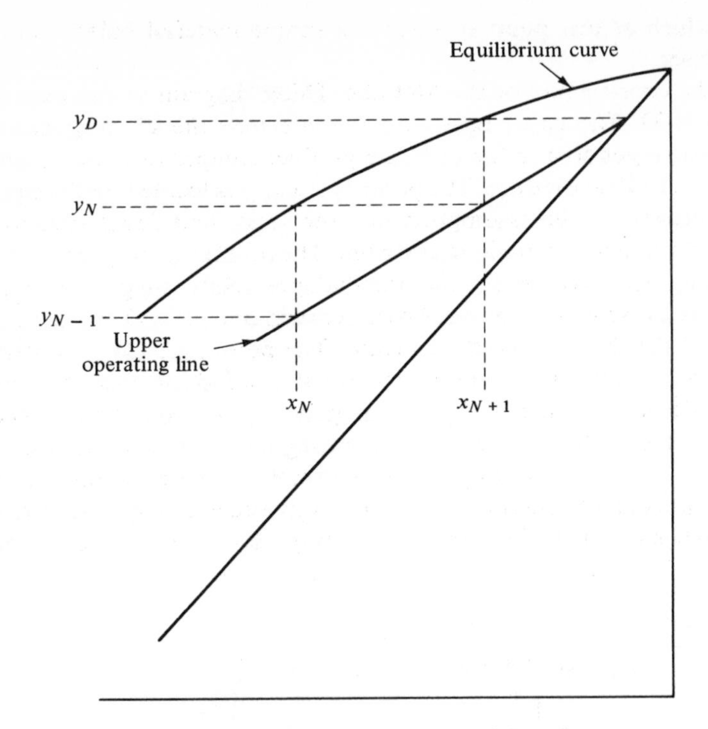

Fig. 4-33. McCabe–Thiele step for a partial condenser.

additional theoretical stage of separation which the idealized partial condenser achieves.

4-18. Fractions of theoretical plates

In performing a plate count on a McCabe–Thiele diagram, it is generally not possible to step an integer number of steps from the bottom product composition up to the distillate product composition and have everything matched up exactly. It is customary, therefore, to express the plate count in terms of an integer number of theoretical plates plus an additional fraction of a theoretical plate. While the method of calculating the number of theoretical plates is somewhat arbitrary in this case, a commonly employed method is illustrated in Fig. 4-34. Assuming that the McCabe–Thiele diagram construction is started at the bottom of the column with the bottom product composition, the steps are carried up the diagram until finally a point is reached where adjacent values of y bracket the required distillate product composition. Referring to Fig. 4-34, let the point (y_{n-1}, x_n) be the last point on the upper operating line, and assume that y_n, which is in equilibrium with x_n, exceeds the required distillate product composition. If plate n were the

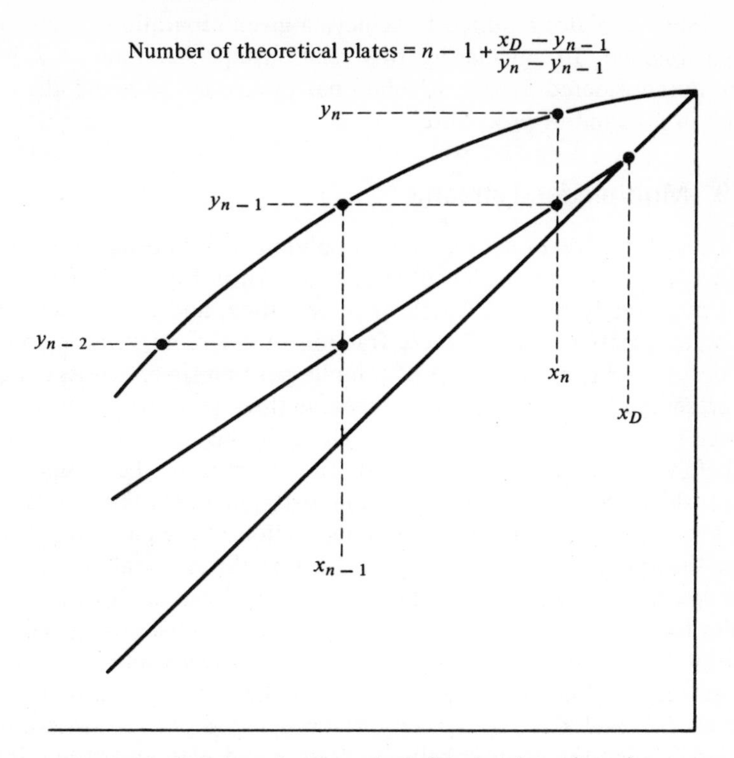

Fig. 4-34. Fraction of a theoretical plate (total condenser).

top plate, the vapor leaving the top plate would be richer than the required distillate product composition; therefore, the required number of plates is fewer than n (this discussion assumes the use of a total condenser). The fraction of the nth plate required is computed as $(x_D - y_{n-1})$ divided by $(y_n - y_{n-1})$, which is the fraction of the last vertical step required to get up to the specified distillate product composition. This fraction is then added to $n - 1$ to yield the total number of theoretical plates required.

This procedure is somewhat arbitrary. One could begin the plate count at the top of the column, starting with the distillate composition and working down, and then computing a fractional plate at the bottom of the column. It is likely that the two plate counts, computed the two different ways, would not agree exactly, but the difference between them would generally be small, and one could surely not be assumed to be correct and the other incorrect.

Naturally, a real distillation column must contain an integer number of actual plates. But some kind of a plate efficiency must be adopted to relate actual plates to theoretical plates, and thus there is no rigorous compulsion to avoid considering a fraction of a theoretical plate. Furthermore, in the optimization of a distillation process, it is desirable to determine a graph of the number of plates required to achieve a given separation versus the reflux ratio employed. In calculating this relationship, fractions of a plate are generally considered in the calculational procedure to avoid the error involved in a roundoff procedure.

4-19. Multiple feed streams

Consider a fractional-distillation column with two feed streams added at different points within the column, as shown in Fig. 4-35. The lower feed is fed at a steady rate of F moles per unit time, and its composition is z_F, which represents its overall mole fraction of the more volatile component. The upper feed is fed at a rate of G moles per unit time, and its composition is represented by z_G. The column contains three sections as far as material-balance relationships are concerned. The stripping section is that section of the column beneath the lower feed. In this section the material-balance relationships are given by Eqs. (4-9) and (4-10), just as in the case of a column with a single feed stream. The enriching section of the column is the section above the upper feed. In the enriching section the material-balance relationships are given by Eqs. (4-11) and (4-12), except that in this case Eq. (4-12) applies for $n = g, g + 1, \ldots, N$. In the column shown in Fig. 4-35, a third material-balance section exists in the region of the column between the two feed plates. In this section of the column the liquid overflow rate will be denoted L'', and the vapor rate will be denoted V''. A material-balance envelope slicing the column between plate p and plate $p + 1$ and encircling the bottom of the column will also intersect the lower feed stream and the bottom product stream. The overall material balance is written

$$L'' + F = V'' + B \tag{4-44}$$

and a material balance for the more volatile component is

$$L''x_{p+1} + Fz_F = V''y_p + Bx_B \qquad (\text{for } p = f, f + 1, \ldots, g - 1) \tag{4-45}$$

Material balances around the entire column are

$$F + G = D + B \tag{4-46}$$

$$Fz_F + Gz_G = Dx_D + Bx_B \tag{4-47}$$

The liquid and vapor flow rates in the various sections are related by

$$L'' - L = q_G G \tag{4-48}$$

$$V - V'' = (1 - q_G)G \tag{4-49}$$

$$L' - L'' = q_F F \tag{4-50}$$

$$V'' - V' = (1 - q_F)F \tag{4-51}$$

where q_G and q_F represent the thermal conditions of the upper and lower feed streams, respectively.

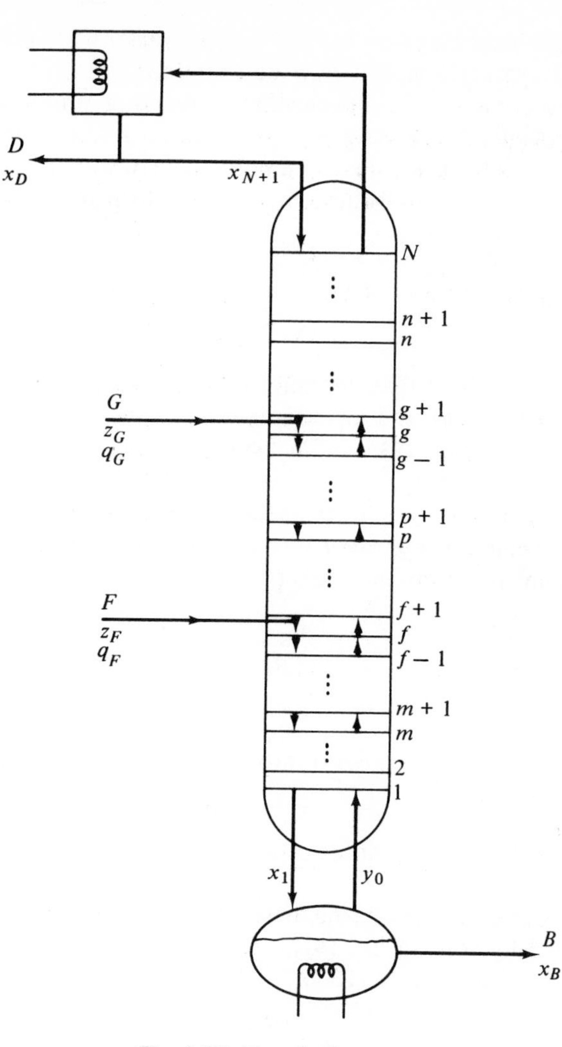

Fig. 4-35. Two feed streams.

The slope of the operating line in the enriching section, the upper operating line, is equal to L/V, as in the case of a column with a single feed stream. Likewise, the upper operating line intersects the 45° diagonal line at a value of x corresponding to x_D. Similarly, for the stripping section, the lower operating line has a slope L'/V' and an intersection with the 45° diagonal at $x = x_B$, just as in the case of a column with a single feed stream. The operating line for the intermediate section, the section of the column between the two feed streams, is given by Eq. (4-45). The slope of this operating line on a McCabe–Thiele diagram is L''/V''. The various intersections of the intermediate operating line will now be examined.

The intersection of the intermediate operating line with the lower operating line occurs along the q line for the lower feed stream. This can be demonstrated by solving Eq. (4-10) simultaneously with Eq. (4-45) for the intersection point (y_i, x_i) by subtracting Eq. (4-10) from Eq. (4-45) to give

$$(L'' - L')x_i = (V'' - V')y_i - Fz_F \tag{4-52}$$

Substituting Eqs. (4-50) and (4-51) into Eq. (4-52) and dividing by F yields

$$-q_F x_i = (1 - q_F)y_i - z_F \tag{4-53}$$

It is seen from Eq. (4-53) that the point representing the intersection of the lower operating line with the intermediate operating line lies on the q line for the lower feed stream, a line with a slope $-q_F/(1 - q_F)$ which intersects the 45° diagonal at $x = z_F$.

The intersection of the intermediate operating line with the upper operating line occurs along the q line for the upper feed. Solving for the intersection point by combining Eqs. (4-12) and (4-45) yields

$$(L'' - L)x_i + Fz_F = (V'' - V)y_i + Bx_B + Dx_D \tag{4-54}$$

Subtracting Eq. (4-47) from Eq. (4-54) yields

$$(L'' - L)x_i = (V'' - V)y_i + Gz_G \tag{4-55}$$

Substituting Eqs. (4-48) and (4-49) into Eq. (4-55) and dividing by G yields

$$q_G x_i = -(1 - q_G)y_i + z_G \tag{4-56}$$

from which it is clear that the intersection point lies on the q line for the upper feed stream.

The intersection of the intermediate operating line with the 45° diagonal line is obtained by solving Eq. (4-45) simultaneously with the equation of the diagonal line, $y = x$, to give

$$x_i = y_i = \frac{Bx_B - Fz_F}{L'' - V''} \tag{4-57}$$

Employing Eq. (4-44), Eq. (4-57) can be rewritten as

$$x_i = y_i = \frac{Bx_B - Fz_F}{B - F} \tag{4-58}$$

The numerator on the right-hand side of Eq. (4-58) is the flow rate of the more volatile component in the bottom product stream minus the flow rate of that component in the lower feed stream; thus the numerator equals the net flow rate of the more volatile component *from* the column in all feed and product streams *beneath* any point in the intermediate section of the column. The denominator on the right-hand side of Eq. (4-58) is simply the bottom product rate minus the feed rate of the lower feed stream, and thus it equals the net flow rate of total moles *leaving* the column in all feed and product streams beneath any point within the intermediate section. Therefore, the right-hand side of Eq. (4-58) is the ratio of the net flow rate of the more volatile component from the column to the net flow of total moles from the column in all feed and product streams beneath any point within the intermediate section. It is thus equal to the mole fraction of the more volatile component in a fictitious stream leaving the bottom of the column which represents the net effect of all feed and product streams beneath the intermediate section in terms of the rate of flow of total moles and the rate of flow of moles of the more volatile component leaving the column.

Combining Eqs. (4-46) and (4-47) with Eq. (4-58) yields

$$x_i = y_i = \frac{Gz_G - Dx_D}{G - D} \tag{4-59}$$

Thus it is seen that the point of intersection of the intermediate operating line with the 45° diagonal line is also represented by the ratio of the net moles of the more volatile component *entering* the column in all feed and product streams *above* any point in the intermediate section divided by the net total moles *entering* the column in all feed and product streams *above* any point within the intermediate section. Needless to say, the sign of the numerator and the sign of the denominator on the right-hand side of Eqs. (4-58) or (4-59) could both be changed, in which case one would think of the composition of a fictitious stream *entering* the bottom of the column or *leaving* the top of the column.

For the particular case in which the net total moles leaving or entering the column in all feed and product streams beneath the intermediate section is equal to zero, the intermediate operating line will intersect the 45° diagonal line at infinity and will therefore be parallel to it.

Finally, it is useful to note that the upper operating line and the lower operating line intersect each other along a q line for the combined feed streams. Combining Eqs. (4-10) and (4-12), the intersection point for the upper and lower operating lines is given by

$$(V - V')y_i = (L - L')x_i + Dx_D + Bx_B \tag{4-60}$$

Adding Eqs. (4-48) and (4-50) gives

$$L' - L = q_F F + q_G G \tag{4-61}$$

and adding Eqs. (4-49) and (4-51) gives

$$V - V' = (1 - q_F)F + (1 - q_G)G \tag{4-62}$$

Substituting Eqs. (4-47), (4-61), and (4-62) into Eq. (4-60) and dividing through by $(F + G)$ yields

$$\frac{(1 - q_F)F + (1 - q_G)G}{F + G} y_i = -\frac{q_F F + q_G G}{F + G} x_i + \frac{F z_F + G z_G}{F + G} \tag{4-63}$$

If the two feed streams were to be combined, the overall mole fraction of the more volatile component in the combined stream would be exactly equal to the last term on the right-hand side of Eq. (4-63). Similarly, the liquid fraction of the combined feed streams would be equal to the negative of the coefficient of x_i on the right-hand side of Eq. (4-63). Thus the fictitious stream obtained by combining the two feed streams would have a composition z_{mix} and a liquid fraction q_{mix}. In terms of these quantities Eq. (4-63) becomes

$$(1 - q_{mix})y_i = -q_{mix}x_i + z_{mix} \tag{4-64}$$

Therefore, the intersection point for the lower and upper operating lines lies on the q line for the fictitious stream obtained by mixing the two feed streams.

When two feed streams are fed to the distillation column, the condition of minimum reflux will generally correspond to a pinch between the operating lines and the equilibrium curve at the intersection of the equilibrium curve with the q line for one of the two feeds. In calculating the minimum reflux ratio, it is important to check to see along which q line the pinch will occur first as the reflux ratio is continuously lowered. Of course, a tangent pinch can also determine the condition of minimum reflux when the equilibrium curve has a point of inflection.

The location of the feed plate for the introduction of a feed stream determines when the steps on a McCabe–Thiele diagram shift from one operating line to the other, just as in the case of a single feed stream. But when two feed streams are employed in a distillation column, the *order* of introduction of the feed streams determines the *position* of the intermediate operating line. Therefore, in this case, feed-stream location can affect the position of the intermediate operating line as well as the position of the steps in the McCabe–Thiele diagram. These concepts are best illustrated by means of an example problem.

EXAMPLE PROBLEM 4-8 Repeat Example Problem 4-5 but for a distillation column to be fed with two feed streams. One feed stream will be a saturated liquid stream containing 30 mole per cent acetone and fed at a rate of 60 lb moles/hr, and the other feed stream will be a saturated vapor containing 92.5 mole per cent acetone and fed at a rate of 40 lb moles/hr.

Solution: If these two feed streams were mixed together, the mixture would be identical with the feed stream of Example Problem 4-5. The mixture would be 60 per cent liquid and 40 per cent vapor, and thus $q_{mix} = 0.6$. The total number of moles per hour fed to the column is 60 + 40, or 100 moles/hr, and the total molal flow rate of acetone fed in the two streams is

$$(0.3)(60) + (0.925)(40) = 55 \text{ lb moles/hr}$$

Therefore, the composition of the mixed feed, z_{mix}, is equal to 0.55. The distillate and bottom product flow rates and compositions will therefore be the same as in Example Problem 4-5. Since the reflux ratio is the same, the upper and lower operating lines will also be the same as in Example Problem 4-5. The McCabe–Thiele diagram, Fig. 4-36, shows the q line for the mixed feed stream and the upper and lower operating lines to be identical with those in Example Problem 4-5. The q line for the liquid feed stream in this problem is a vertical line at $x = 0.3$. The q line for the vapor feed stream in this problem is a horizontal line at $y = 0.925$.

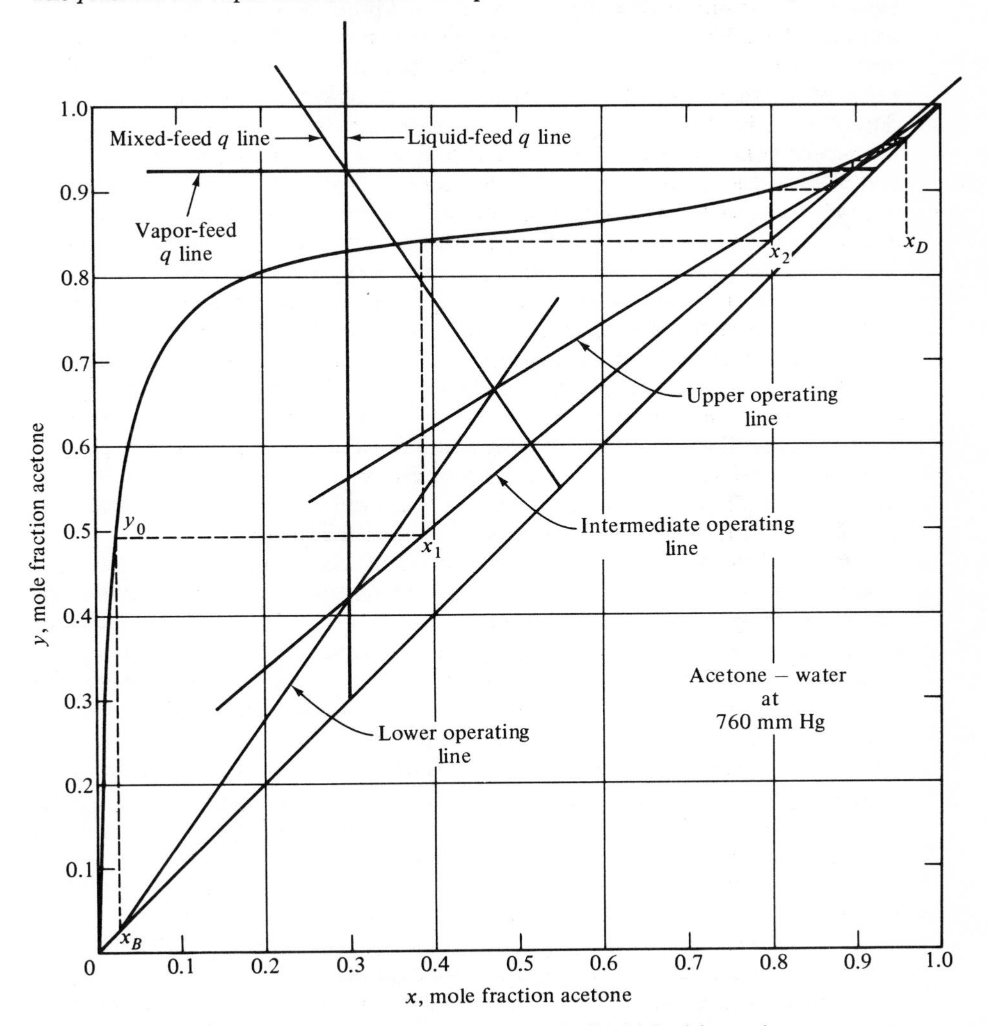

Fig. 4-36. Solution to Example Problem 4-8; liquid feed beneath vapor feed.

Note that these two q lines intersect each other at the same point at which they intersect the q line for the mixed feed stream.

At this point a choice must be made regarding the order of introduction of the two feed streams. If the liquid feed stream is introduced beneath the vapor feed stream, then the intermediate operating line will intersect the lower operating line along the q line for the liquid feed stream, and the intermediate operating line will intersect the upper operating line along the q line for the vapor feed stream. With this choice for the order of introducing the feed streams, the intermediate operating line will take the position shown in Fig. 4-36.

The theoretical plates are then stepped off, with the change from one operating line to the next one being made as soon as it will produce a higher rate of advancement up the column. The steps are also shown in Fig. 4-36. The point (y_0, x_1) is seen to lie on the intermediate operating line, because this choice results in a higher value of x_1 than that which would result if the lower operating line were used. Since this point lies on the intermediate operating line, the liquid feed must be introduced beneath the first plate. It can be mixed with the liquid overflow from the first plate to the reboiler, or it can be fed directly to the reboiler; if the reboiler is assumed to be well-mixed, the result will be the same in either case. Continuing up the column, the last point on the intermediate operating line is the point (y_3, x_4). The point (y_4, x_5) is located on the upper operating line because this choice yields a larger value of x_5 than would be obtained if the intermediate operating line were to be used. Therefore, the vapor feed is to be fed to the fourth plate (from beneath). Continuing the count up the column, approximately seven theoretical plates plus the equilibrium rebolier are required. Actually, the number of theoretical plates required is lower than the number required in Example Problem 4-5. The use of the intermediate operating line allows the steps to advance the composition faster than the steps in Example Problem 4-5. But the accuracy of the graphical constructions shown for the two example problems does not permit a determination of the difference between the numbers of theoretical plates required in the two cases.

It would be possible to feed the vapor feed stream beneath the liquid feed stream. In this case, the intersection of the intermediate operating line with the lower operating line would occur along the q line for the vapor feed, and the intersection of the intermediate operating line with the upper operating line would occur along the q line for the liquid feed stream. Thus the intermediate operating line would take the position shown in Fig. 4-37. Shown also on the diagram are some steps. The point (y_0, x_1) is located on the lower operating line, but the point (y_1, x_2) is located on the intermediate operating line. Thus the *vapor* feed stream is fed to the bottom plate (from beneath). For the steps shown, it is seen that the point (y_2, x_3) is located on the upper operating line; therefore, the liquid feed is fed to plate 2. Stepping on up the column, approximately eight theoretical plates are required.

Of course, more steps could have been taken on the intermediate operating line before shifting to the upper operating line, but it is clear from the diagram that these steps would have simply stepped into the pinch point and would have made little progress in terms of increasing the acetone concentration. Actually, the number of theoretical plates required to achieve the desired separation would be decreased by not taking any steps on the intermediate operating line but simply stepping from the point (y_0, x_1) on the lower operating line to the point (y_1, x_2) on the upper operat-

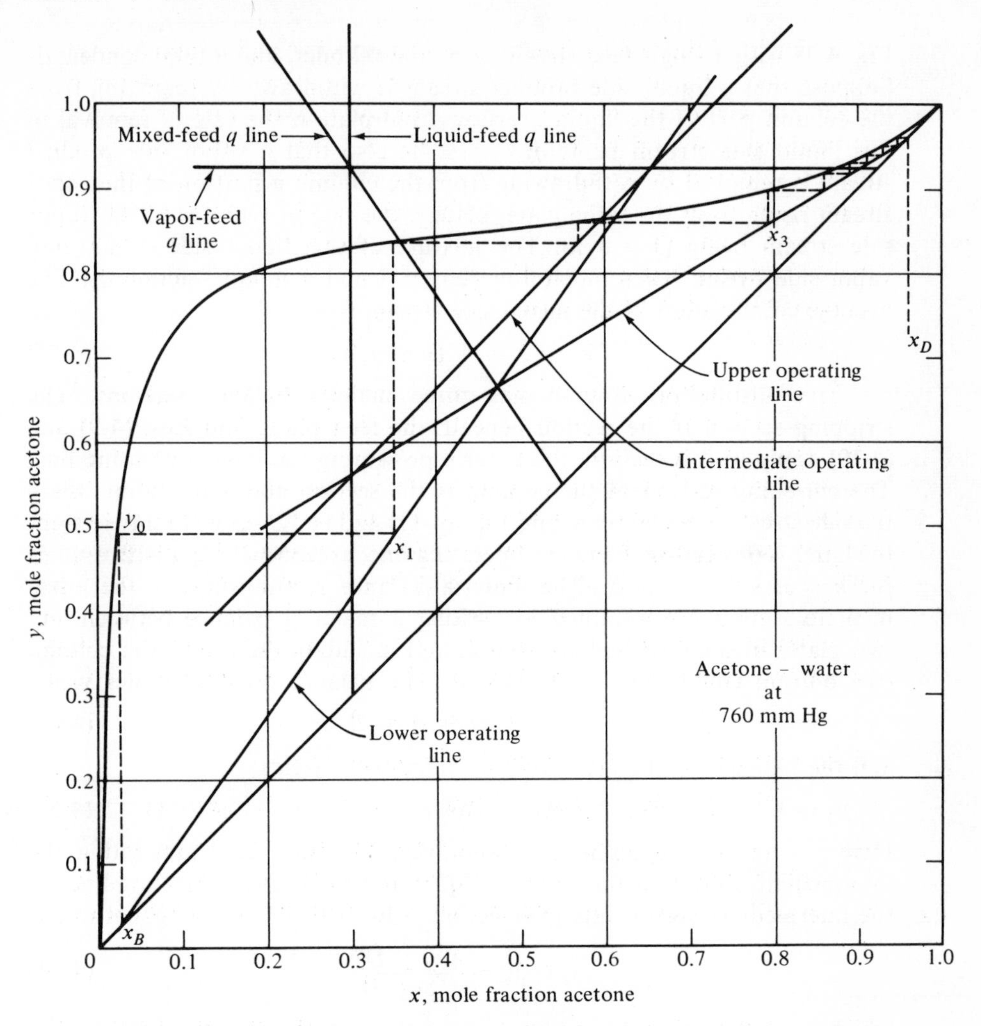

Fig. 4-37. Alternative solution for Example Problem 4-8; vapor feed beneath liquid feed.

ing line. This would correspond to feeding both the vapor feed and the liquid feed to the first plate, and the result would be the same as in Example Problem 4-5. But the best way to arrange the feed streams corresponds to Fig. 4-36, that is, feeding the liquid feed into the liquid reboiler and the vapor feed to the fourth plate.

4-20. Side product streams

It is sometimes desirable to remove side product streams from the middle of a distillation column. For example, consider the column shown in

Fig. 4-38 with a single feed stream, a partial reboiler, and a total condenser. Suppose that a liquid side product stream is withdrawn by removing from the column part of the liquid overflow from plate s, the rate of removal of this liquid side stream being $q_S S$. Assume also that a vapor side product stream is removed by withdrawing from the column a portion of the vapor stream rising from the same plate, plate s, the rate of removal of the vapor side stream being $(1 - q_S)S$. The mixture of this liquid side stream and vapor side stream has a molal flow rate of S and a liquid fraction q_S. The average composition of the mixed side stream is

$$z_S = q_S x_s + (1 - q_S)y_s \tag{4-65}$$

This distillation column has three material-balance sections. The stripping section is the section beneath the feed plate, and Eqs. (4-9) and (4-10) apply to this section, the latter representing the lower operating line. The enriching section of the column is the section above the point where the side stream is withdrawn, and Eqs. (4-11) and (4-12) apply to this section, the latter representing the upper operating line, except that Eq. (4-12) applies for $n = s, s + 1, \ldots, N$. The material-balance relationships in the intermediate section are obtained by writing a material balance between any two plates in the intermediate section and including the top of the column (the bottom could have been chosen). The balance on total moles yields

$$V'' = L'' + S + D \tag{4-66}$$

and the balance on the more volatile component yields

$$V''y_p = L''x_{p+1} + Sz_S + Dx_D \qquad (\text{for } p = f, f + 1, \ldots, s - 1) \tag{4-67}$$

Here V'' and L'' represent the vapor rate and the liquid rate within the intermediate section of the column. Equation (4-67) is the operating line for the intermediate section. Its intersection point with the 45° diagonal line is

$$x_i = y_i = \frac{Sz_S + Dx_D}{S + D} \tag{4-68}$$

which is simply the composition of a fictitious stream which represents the net flow out of the column from all feed and product streams above any point within the intermediate section.

The intersection of the intermediate operating line and the upper operating line lies on the q line for the side stream. Solving Eqs. (4-12) and (4-67) simultaneously for the intersection point yields

$$(V'' - V)y_i = (L'' - L)x_i + Sz_S \tag{4-69}$$

But the changes in the liquid and vapor flow rates that result from the removal of the side stream are

$$V'' - V = (1 - q_S)S \tag{4-70}$$

$$L - L'' = q_S S \tag{4-71}$$

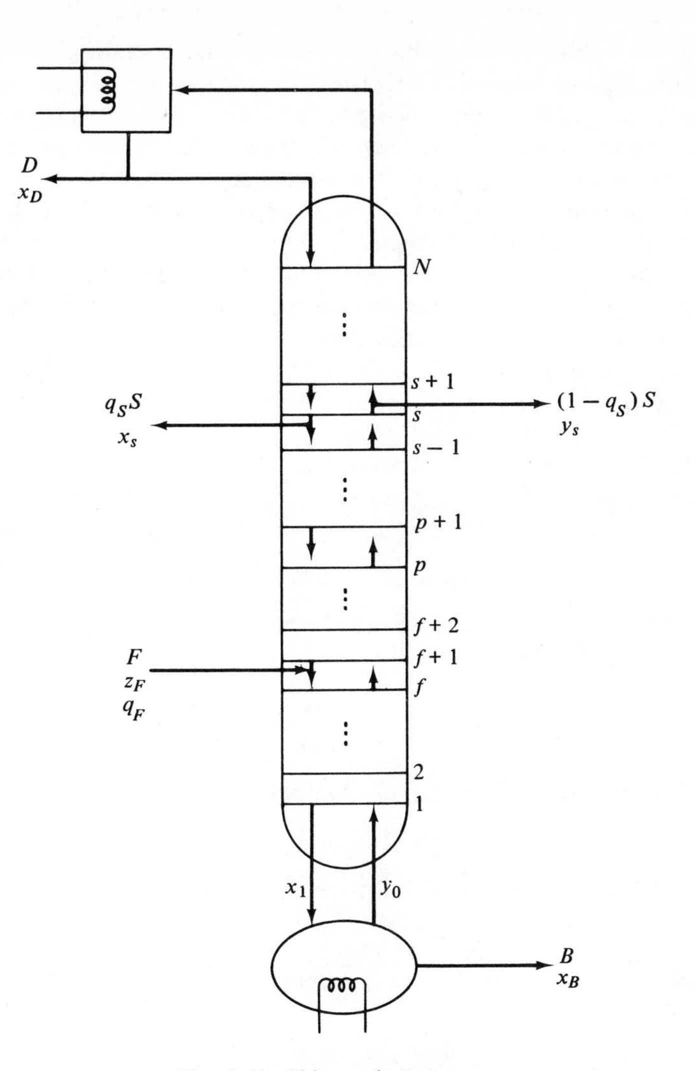

Fig. 4-38. Side product stream.

Inserting Eqs. (4-70) and (4-71) into Eq. (4-69) and dividing through by S yields

$$(1 - q_S)y_i = -q_S x_i + z_S \tag{4-72}$$

It is seen, therefore, that the intersection of the upper operating line and the intermediate operating line lies on the q line for the side stream, where this q line has a slope $-q_S/(1 - q_S)$ and intersects the 45° diagonal at z_S. It will be left to the reader to show that the intermediate operating line and the lower operating line intersect along the q line for the feed stream.

Thus it is apparent that a side product stream has the same properties as a feed stream in regard to q lines and the intersection of operating lines in adjacent sections separated by the introduction of a feed stream or the withdrawal of a side stream. One difference appears with respect to the change in slope of the operating line. When two sections of the column are separated by a feed stream, the slope of the operating line will be lower in the section above the feed plate than in the section beneath the feed plate. When two sections of the column are separated by a side product stream, the slope of the operating line in the section above the side product withdrawal point will be higher than the slope of the operating line in the section beneath the side-product-stream withdrawal point.

Referring again to the column shown in Fig. 4-38, the upper operating line, represented by Eq. (4-12), and the lower operating line, represented by Eq. (4-10), intersect at the point (y_i, x_i) which is given by the equation

$$(V - V')y_i = (L - L')x_i + Dx_D + Bx_B \tag{4-73}$$

The introduction of the feed stream makes the following changes in the vapor and liquid rates:

$$L' - L'' = q_F F \tag{4-74}$$

$$V'' - V' = (1 - q_F)F \tag{4-75}$$

Subtracting Eq. (4-70) from Eq. (4-75) yields

$$V - V' = (1 - q_F)F - (1 - q_S)S \tag{4-76}$$

and subtracting Eq. (4-74) from Eq. (4-71) yields

$$L - L' = q_S S - q_F F \tag{4-77}$$

The material balance on the more volatile component around the entire column is

$$Fz_F = Dx_D + Bx_B + Sz_S \tag{4-78}$$

Inserting Eqs. (4-76), (4-77), and (4-78) into Eq. (4-73) and dividing by $(F - S)$ yields

$$\frac{(1 - q_F)F - (1 - q_S)S}{F - S} y_i = -\frac{q_F F - q_S S}{F - S} x_i + \frac{Fz_F - Sz_S}{F - S} \tag{4-79}$$

The last term on the right-hand side of Eq. (4-79) is the composition of a fictitious feed stream which introduces the same number of total moles and the same number of moles of the more volatile component into the column as corresponds to the net effect of *introducing* the feed stream and *withdrawing* the side stream. Therefore, this term represents the composition of a "mixture" of the feed stream and the side product stream, which really represents the feed stream *minus* the side product stream. The coefficient of x_i in Eq. (4-79) is the negative of the liquid fraction of a fictitious feed stream which introduces the same number of total moles and the same number of moles of liquid into the column as the net effect of adding the feed stream and withdrawing the side product stream. Similarly, the coefficient of y_i in Eq. (4-79) corresponds to the vapor fraction of a fictitious feed stream which introduces the same number of total moles and the same number of moles of vapor into the column as the net effect of introducing the feed stream and withdrawing the side product stream. Denoting these quantities as z_{mix}, q_{mix}, and $1 - q_{\text{mix}}$, respectively, Eq. (4-79) can be written as

$$(1 - q_{\text{mix}})y_i = -q_{\text{mix}}x_i + z_{\text{mix}} \tag{4-80}$$

Thus the upper and lower operating lines for the column shown in Fig. 4-38 intersect along a q line which represents the "mixture" of the feed stream and the side product stream, with this mixture stream being a fictitious feed stream which would *introduce* into the column the same number of total moles, moles of liquid, moles of vapor, and moles of more volatile component as the net effect of *adding* the feed stream and *withdrawing* the side product stream.

A section of a distillation column for which a straight operating line represents the material balance consists of any portion of the column into which no feed streams are introduced and from which no side product streams are withdrawn. The introduction of a feed stream or the withdrawal of a side product stream will correspond to a boundary between adjacent sections. The operating lines describing the material-balance relationships in any two adjacent sections will intersect along the q line corresponding to the feed or side product stream that separates the two sections. These simple concepts are generalized in a quite straightforward manner to include any number of feed streams and any number of side product streams, arranged in any order.

EXAMPLE PROBLEM 4-9 A saturated vapor feed stream containing 70 mole per cent benzene and 30 mole per cent toluene is to be fed continuously at a rate of 100 lb moles/hr to a fractional-distillation column operating at a pressure of 1 atm. The column will be equipped with a partial reboiler and a total condenser. It is desired to produce a side product stream at a rate of 20 lb moles/hr. The side product stream will be a mixture of 50 per cent liquid and 50 per cent vapor and will have an average composition of 30 mole per cent benzene. A distillate product stream which is 95 mole per cent benzene and a bottom product stream which is 5 mole per cent benzene will also be produced from the column. If a reflux ratio of 1.5 is employed, how many theoretical plates are required to achieve this separation? To which plate should the feed stream be added and from which plate should the side product stream be removed? What must be the molal rate of generation of vapor within the reboiler?

Solution: The material balance on total moles for the entire column is

$$D + B = F - S = 100 - 20 = 80$$

and the balance on benzene is

$$(0.95)D + (0.05)B = (100)(0.7) - (20)(0.3) = 64$$

Solving these two equations simultaneously yields $D = 66.67$ and $B = 13.33$.

With the reflux ratio of 1.5, the slope of the upper operating line is

$$\frac{L}{V} = \frac{1.5}{2.5} = 0.6$$

Thus the upper operating line is located as a line with a slope of 0.6 which intersects the 45° diagonal line at a composition corresponding to 95 mole per cent benzene.

The properties of the fictitious "mixture" feed stream which represents the net effect of adding the feed stream and withdrawing the side product stream are

$$z_{\text{mix}} = \frac{(100)(0.7) - (20)(0.3)}{100 - 20} = 0.8$$

$$q_{\text{mix}} = \frac{(100)(0) - (20)(0.5)}{100 - 20} = -\frac{10}{80} = -\frac{1}{8}$$

The slope of the q line for this fictitious mixture stream is

$$\frac{-q_{\text{mix}}}{1 - q_{\text{mix}}} = \frac{\frac{1}{8}}{1 + \frac{1}{8}} = \frac{1}{9}$$

Therefore, the q line for the fictitious mixture stream is located as a line with a slope of $\frac{1}{9}$ which intersects the 45° diagonal at a composition corresponding to 80 mole per cent benzene. The upper and lower operating lines intersect along this q line for the mixture stream, and the lower operating line also intersects the 45° diagonal line at a composition corresponding to 5 mole per cent benzene. Therefore, the position of the lower operating line is fixed by these two intersections. These lines are shown in Fig. 4-39.

The q line for the side product stream has a slope of -1, and it intersects the 45° diagonal line at a composition corresponding to 30 mole per cent benzene. This q line is readily located on the McCabe–Thiele diagram. *Assuming* that the side product stream will be withdrawn from a plate *beneath* the feed plate, the intermediate operating line will intersect the lower operating line along the q line for the side product stream and will intersect the upper operating line along the q line for the feed stream, which is a horziontal line corresponding to $y = 0.7$. This locates the position of the intermediate operating line.

As a check, the intersection of the intermediate operating line with the 45° diagonal will be computed. Looking to the bottom of the column, the side product stream and the bottom product stream both leave the column beneath the intermediate section of the column. The composition corresponding to a mixture of these two streams is

$$\frac{(20)(0.3) + (13.33)(0.05)}{20 + 13.33} = 0.2$$

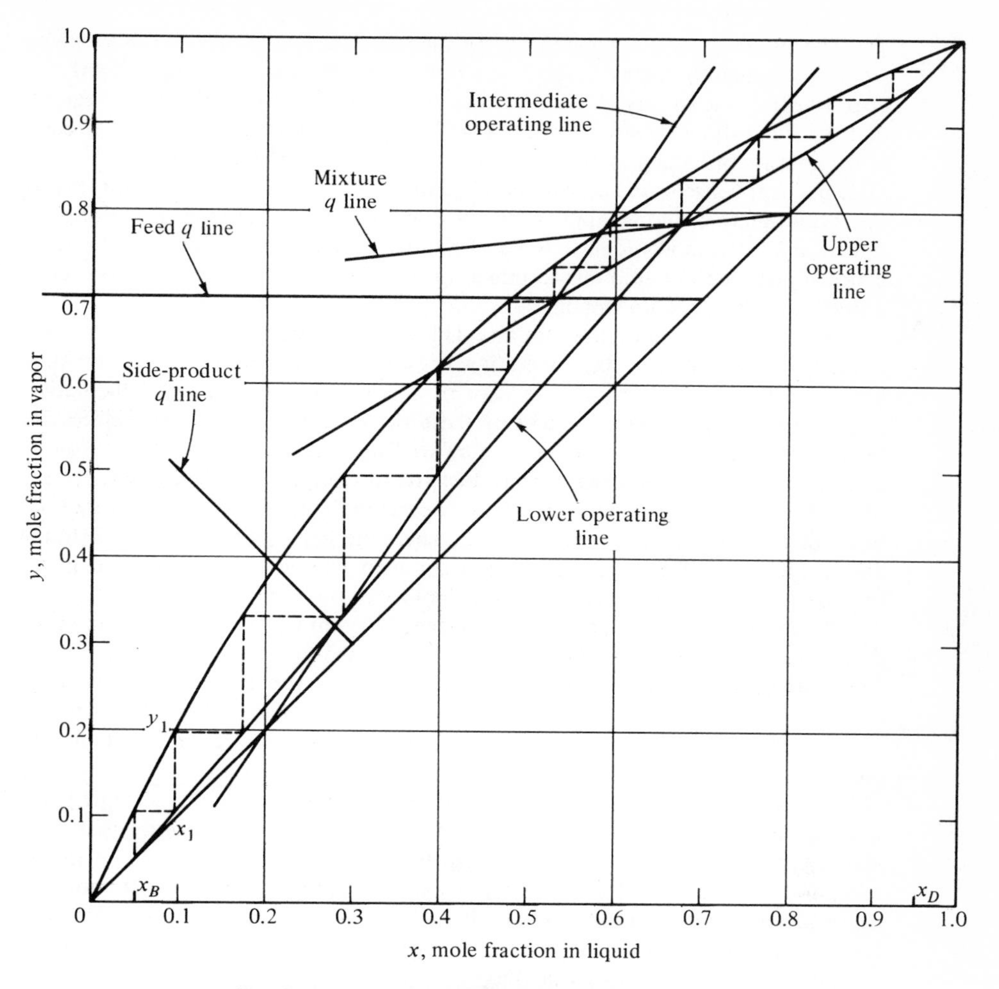

Fig. 4-39. Solution to Example Problem 4-9.

and it can be seen in Fig. 4-39 that the intermediate operating line does indeed intersect the 45° diagonal line at this composition.

The steps are shown in Fig. 4-39. The point (y_0, x_1) lies on the lower operating line, and so does the point (y_1, x_2). Referring to the diagram, it would surely be advantageous if the intermediate operating line could be used, because it is farther away from the equilibrium curve in this region, and its use would result in a more rapid advancement of the composition up the column. But it is not possible to switch over to the intermediate operating line until after the side stream has been withdrawn, and the side stream cannot be withdrawn from the first or second plates because the benzene mole fractions of the liquid and the vapor on these plates are too low to make the required composition of the side product stream.

The point (y_2, x_3) is located on the intermediate operating line (although x_3 would be negligibly higher if the lower operating line were used because the two operating lines are so close to each other at this point). Strictly speaking, this requires that the side product sream be withdrawn from the second plate. Actually, an equimolal mixture of the liquid and the vapor leaving the third plate would be about 39 mole per cent benzene and an equimolal mixture of the liquid and the vapor leaving the second plate would be approximately 25 mole per cent benzene. Therefore, the side product stream would have to be withdrawn from both the second and the third plates in just the right amounts to make up the 30 mole per cent benzene composition for the side product stream.

Note that the steps on the McCabe–Thiele diagram step around the intersection of the lower and intermediate operating lines always stepping onto the operating line *nearer* the equilibrium curve so as to produce a *maximum* number of theoretical plates for stepping up the column using these two operating lines. This is necessary because the side stream must be withdrawn from the plate with a composition corresponding to the required side product stream composition. This is the opposite of the situation at an intersection of two operating lines representing two sections of the column separated by a feed stream, where the steps are generally made to the line farther away from the equilibrium curve to *minimize* the number of plates required to step up this section of the column.

Proceeding up the column, it is seen that the point (y_5, x_6) lies on the intermediate operating line but the point (y_6, x_7) lies on the upper operating line. Therefore, the feed stream is fed to the sixth plate. Up at the top it is seen that the vapor leaving the eleventh plate is richer in benzene than the required distillate product stream; therefore, approximately 10.5 theoretical plates are required in the column.

In order for the side stream to be withdrawn as specified from a single plate, it would be necessary that the liquid and vapor compositions leaving that plate should correspond exactly to the intersection of the side product q line with the equilibrium curve. Thus for the present example, the liquid would be 21 mole per cent benzene and the vapor would be 39 mole per cent benzene. An equimolal mixture of that liquid and that vapor would contain 30 mole per cent benzene. If the plate count on a McCabe–Thiele diagram is begun with the bottom product composition, it is very improbable that any step would land exactly in the right place so that the liquid and vapor compositions on a given plate would correspond exactly to the intersection of the side product q line with the equilibrium curve. Of course, the McCabe–Thiele construction could be started at the plate from which the side product is to be withdrawn, locating the intersection of the side product q line and the equilbirium curve and calling that the point (y_s, x_s). Steps could then be taken up and down the column from that point, and in all probability a fraction of a theoretical plate would be required at the bottom of the column and also at the top of the column. Of course, if two side product streams are withdrawn from the column, such a procedure could not make them both come out exactly right. Furthermore, it would be prudent in building the actual column to locate pipes for withdrawing the side product stream at several adjacent trays in order that the operating column can be made to produce a side product of the right composition.

Therefore, the construction shown in Fig. 4-39 is probably adequate. It simply shows that the side product withdrawn from the second plate would be too rich in toluene while the side product withdrawn from the third plate would be too rich in benzene, and therefore a mixture of streams from these two plates will be required.

4-21. Column and plate efficiencies

The plates of a real distillation column will not, in general, bring the vapor and liquid phases leaving the plate to equilibrium with each other; therefore, an efficiency is defined to relate the actual distillation-column performance to the performance of a column composed of theoretical plates. The simplest type of efficiency in common usage is the overall column efficiency, defined by Lewis [3] as the ratio of the number of theoretical plates to the number of actual plates required to effect the same separation at the same reflux ratio.

$$\mathscr{E}_{\text{col}} \equiv \left(\frac{\text{NTP}}{\text{NAP}}\right)_{\text{same separation}} \tag{4-81}$$

Using an overall column efficiency in the design of a distillation column is, therefore, quite straightforward. After determining the number of theoretical plates required for the separation, this number is simply divided by the column efficiency to obtain the actual number of plates required.

Although the overall column efficiency is widely used, the Murphree [5] plate efficiency is preferred because of its more solid theoretical basis. Murphree plate efficiency is defined as the change in the vapor composition in passing through a plate divided by the change that would have occurred if the vapor had been brought to equilibrium with respect to the liquid composition corresponding to the liquid actually leaving the plate:

$$\mathscr{E}_{\text{plt}} \equiv \frac{y_n - y_{n-1}}{y_n^* - y_{n-1}} \tag{4-82}$$

Here y_{n-1} and y_n refer to the actual compositions of the vapor streams entering and leaving the nth plate, respectively. The symbol y_n^* refers to the vapor composition which corresponds to equilibrium with x_n, which is the actual composition of the liquid stream leaving the nth plate.

When Murphree plate efficiency is employed in the design of a distillation column the actual stepping off of the steps in the McCabe–Thiele diagram is altered. The denominator on the right-hand side of Eq. (4-82) represents the height of the vertical step from the operating line up to the equilibrium line, from the point (y_{n-1}, x_n) up to the point (y_n^*, x_n), as shown in Fig. 4-40. The numerator on the right-hand side of Eq. (4-82) represents the height of the vertical step from the point (y_{n-1}, x_n) on the operating line up to the point (y_n, x_n) which locates the actual composition of the vapor

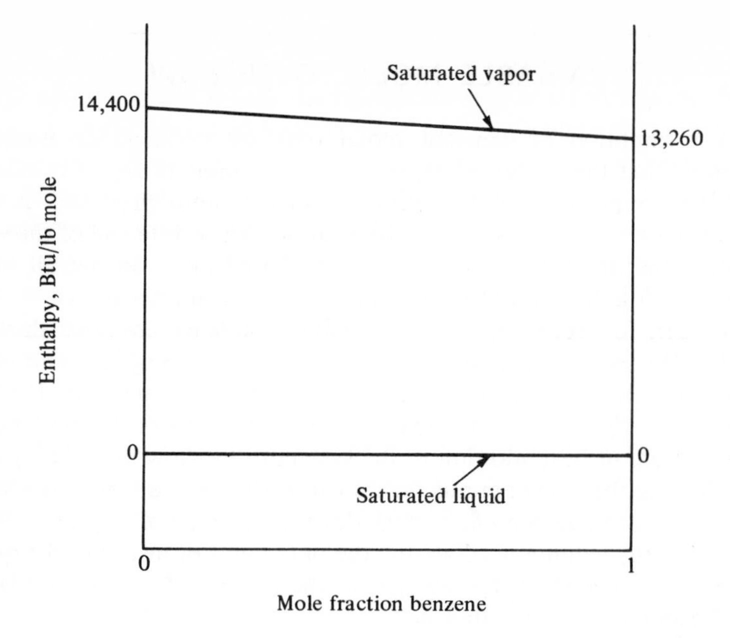

Fig. 4-41. Enthalpy–composition diagram for benzene–toluene at 1 atm.

in which no chemical reactions are occurring, two arbitrary reference conditions may be chosen to define the enthalpy reference point. In Fig. 4-41 the value of zero enthalpy has been arbitrarily (but very conveniently) assigned to pure saturated benzene liquid and to pure saturated toluene liquid. It should be remembered that this curve is for a constant pressure of 1 atm; therefore, the temperature for pure saturated liquid benzene is 80.12°C, the boiling point of benzene, while the temperature corresponding to the point for pure saturated liquid toluene is 110.6°C, the boiling point of toluene. Nevertheless, it is both permissible and convenient to choose for the enthalpy reference points pure liquid benzene at 80.12°C and pure liquid toluene at 110.6°C and to assign zero enthalpy to each of them. With these choices for the reference points, the enthalpy of pure saturated benzene vapor at 80.12°C is equal to 13,260 Btu/lb mole of benzene, which is the enthalpy of vaporization (or *heat of vaporization*) of benzene at its boiling point. Similarly, the enthalpy of pure saturated toluene vapor at 110.6°C is 14,400 Btu/lb mole of toluene because this is the molal latent heat of vaporization of toluene at its normal boiling point. The terminal points of the saturated liquid line are therefore chosen to be zero enthalpy as the arbitrary reference points. The terminal points of the saturated vapor line then automatically become equal to the molal latent heats of vaporization of toluene and of benzene.

The representation of the saturated liquid curve by a straight line through its terminal points is only an approximation. The true curve will depart from the straight line through its terminal points because of the heat of mixing of the liquid and because of sensible heat effects owing to temperature changes as the liquid composition changes. For the benzene–toluene system the heat of mixing is very small and can be neglected. The sensible heat effects are related approximately to the nonlinearity of the bubble-point temperature curve shown in Fig. 4-2. If the bubble-point temperature curve were a straight line between the boiling points of the pure liquids and if the molal heat capacity of benzene were exactly equal to that of toluene, the sensible heat effect would be zero. In actuality it is very, very small.

For example, consider a liquid with a benzene mole fraction of 0.5. The bubble-point temperature at a pressure of 1 atm is read from Fig. 4-2 as 92.5°C. This is 12.4°C warmer than the boiling point of benzene, and since the molal heat capacity of liquid benzene in this temperature range is approximately 37 Btu/lb mole°F, the enthalpy of 0.5 lb mole of liquid benzene at 92.5°C is

$$(0.5 \text{ lb mole})\left(\frac{37 \text{ Btu}}{\text{lb mole }^\circ\text{F}}\right)(12.4^\circ\text{C})\left(\frac{1.8^\circ\text{F}}{^\circ\text{C}}\right) = 413 \text{ Btu}$$

Similarly, 92.5°C is colder than the boiling point of toluene by 18.1°C, and the molal heat capacity of liquid toluene in this temperature range is approximately 44 Btu/lb mole °F; therefore, the enthalpy of 0.5 lb mole of liquid toluene at 92.5°C is

$$(0.5 \text{ lb mole})\left(\frac{44 \text{ Btu}}{\text{lb mole }^\circ\text{F}}\right)(-18.1^\circ\text{C})\left(\frac{1.8^\circ\text{F}}{^\circ\text{C}}\right) = -718 \text{ Btu}$$

Since the reference points for zero enthalpy are at different temperatures for benzene and toluene, the enthalpy of benzene is positive and the enthalpy of toluene is negative at 92.5°C. Neglecting the heat of mixing, the enthalpy of 1 mole of a liquid mixture with a benzene mole fraction of 0.5 is

$$+413 - 718 = -305 \text{ Btu/lb mole of mixture}$$

Thus it is seen that the saturated liquid line dips down to an enthalpy of −305 Btu/lb mole at a benzene mole fraction of 0.5. But 305 is only approximately 2 per cent of 14,000; therefore, the departure of the saturated liquid curve from the straight-line representation is only approximately 2 per cent of the latent heat of vaporization. A similar analysis for the vapor phase will show an even smaller depature from the straight-line relationship, because the molal heat capacities of benzene and toluene vapor are smaller than the liquid heat capacities.

4-23. Variation in the vapor rate throughout the column

Using Fig. 4-41 with the saturated vapor and saturated liquid curves approximated by straight lines through their terminal points, the variation of the molal vapor rate throughout the distillation column is readily understood by referring to Eq. (4-83). With the approximation employed, the enthalpy of any saturated liquid phase is zero, irrespective of its benzene mole fraction. Therefore, Eq. (4-83) simplifies to

$$V'_m H_m = Q_R \tag{4-84}$$

This equation states that the product of the molal vapor rate and the molal enthalpy of the saturated vapor is the same for all plates in the stripping section of the column and is equal to the rate of heat addition to the reboiler. Furthermore, if the feed stream is a saturated liquid, the feed stream will introduce no enthalpy into the column, and Eq. (4-84) would apply to any plate, including the enriching section as well as the stripping section of the column.

For a benzene–toluene distillation column making high-purity distillate and bottom products, the vapor entering the bottom plate would be almost pure toluene, while the vapor leaving the top plate would be almost pure benzene. Thus the molal enthalpy of the vapor stream would vary from 14,400 Btu/lb mole at the bottom of the column to 13,260 Btu/lb mole at the top of the column, a variation of approximately 9 per cent. Equation (4-84) reveals that the molal vapor rate at the top of the column would be approximately 9 per cent higher than the molal vapor rate at the bottom of the column, because of the decrease in the molal enthalpy of the vapor with increasing benzene mole fraction.

It is clear from Fig. 4-41 and Eq. (4-84) that the molal vapor rate would be constant if the molal latent heat of vaporization of benzene were exactly equal to that of toluene and if in addition departures of the saturated liquid and the saturated vapor curves from linearity were truly negligible. It has already been seen that departures from linearity amount to approximately 2 per cent for the saturated liquid line, and they are generally about half as large for the saturated vapor line. But the molal latent heat of vaporization of toluene is 9 per cent larger than the molal heat of vaporization of benzene, and thus a 9 per cent variation in the molal vapor rate will surely result.

Many systems separated by distillation include components whose molal latent heats of vaporization are approximately the same, and thus the assumption of constant molal overflow is a reasonable one. But variations substantially larger than 9 per cent can be found, and in these cases the assumption of constant molal overflow is not acceptable. Consider, for example, the acetone–water system discussed in Example Problems 4-5 through 4-8. In these Example Problems the assumption of constant molal overflow was employed. But the molal heat of vaporization of acetone at its normal boiling point is 13,030 Btu/lb mole, while the molal heat of vaporization of water at its normal boiling point is 17,480 Btu/lb mole, 34 per cent greater than that for acetone. Therefore, in an acetone–water distillation column with a saturated liquid feed and with high-purity product streams, the molal vapor rate at the top of the column will be approximately 34 per cent greater than the molal vapor rate at the bottom of the column.

But while the molal latent heats of vaporization for water and for acetone differ by 34 per cent, the saturated liquid and the saturated vapor curves on the enthalpy–composition diagram depart from linearity only to about the same extent as those for the benzene–toluene system. Therefore, the departures from linearity on the enthalpy–composition diagram result in errors of only about 3 per cent, while the difference in the molal latent heats of vaporization results in errors of 34 per cent in the assumption of constant molal overflow.

4-24. Modified latent heat of vaporization method

These considerations suggest that a substantial improvement in accuracy will result if the assumption of constant molal overflow is replaced by the assumption that the vapor rate is constant when expressed in units of that quantity of acetone and that quantity of water which have the same latent heat of vaporization. For example, instead of considering a mole of acetone with a molecular weight of 58.08, one could consider a *pseudo mole* of acetone with a pseudo molecular weight of $58.08 \times 1.34 = 77.9$. The latent heat of vaporizing 1 pseudo mole of acetone would then become 17,480 Btu/pseudo mole, exactly the same as the heat of vaporization for 1 mole of water. The pseudo molecular weight of water would be taken as 18, the real molecular weight of water. The enthalpy–composition diagram for the acetone–water system would then be plotted in a manner similar to the diagram shown in Fig. 4-41, except that the enthalpy would be the enthalpy per pseudo mole, expressed in Btu per pseudo pound mole of mixture, and the composition would be expressed as a pseudo mole fraction. Approximating the saturated liquid and the saturated vapor curves on the enthalpy–composition diagram as straight lines through their terminal points, Figs. 4-42 and 4-43 show the enthalpy–composition diagram for the acetone–water system at a pressure of 1 atm based upon true molecular weights and pseudo molecular weights, respectively.

Referring to Eq. (4-83), if the vapor and liquid flow rates are expressed in pseudo moles per unit time and the enthalpy quantities are expressed in Btu per pseudo mole of mixture, the form of the equation remains the same.

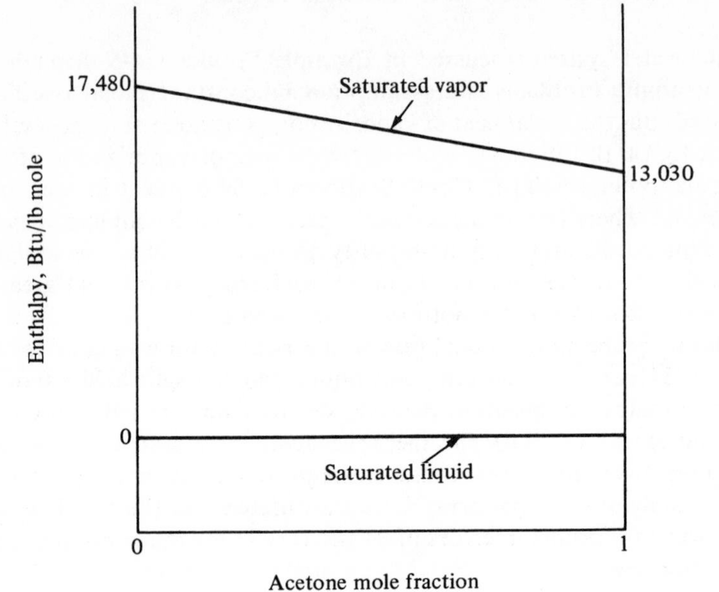

Fig. 4-42. Enthalpy–composition diagram for acetone–water at 1 atm.

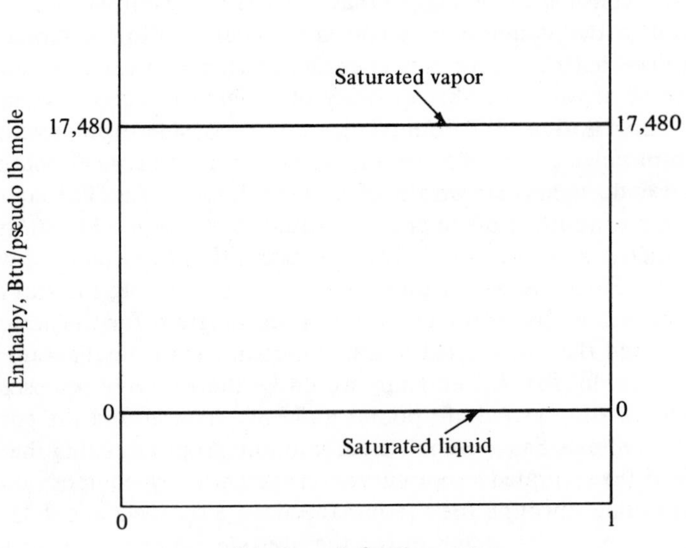

Fig. 4-43. Enthalpy–composition diagram for acetone–water using pseudo molecular weights.

The choice of pure saturated liquids for the enthalpy reference point and the assumption that the saturated liquid curve is a straight line still result in all liquid streams having zero enthalpy, and therefore Eq. (4-84) results as before. The left-hand side of Eq. (4-84) simply expresses the flow rate of enthalpy in the vapor stream, and this can be expressed as the vapor flow rate in pseudo moles flowing per unit time multiplied by the vapor enthalpy in Btu per pseudo mole of mixture. The choice of 77.9 and 18 for the pseudo molecular weights of acetone and water, respectively, is just that choice that makes the latent heat of vaporization of a pseudo mole of acetone equal to the heat of vaporization of a pseudo mole of water; Fig. 4-43 reveals that this choice results in a saturated-vapor enthalpy curve which is a horizontal line of constant enthalpy equal to 17,480 Btu per pseudo mole of mixture, independent of the vapor composition. Equation (4-84) reveals that this choice will therefore result in a vapor flow rate throughout the acetone–water distillation column that is constant when expressed as pseudo moles of vapor flowing per unit time.

The choice of 77.9 and 18 for the pseudo molecular weights of acetone and water, respectively, is by no means the only choice available. Equally satisfactory results could be obtained by choosing 58.08, the true molecular weight, as the pseudo molecular weight for acetone, but then employing $18/1.34 = 13.44$ as the pseudo molecular weight for water. Of course, there are an infinite number of other choices of the pseudo molecular weights for water and for acetone that will result in their latent heats of vaporization per pseudo mole being equal. The term "pseudo mole" is somewhat unfortunate because it suggests a relationship to actual molecular weights, although molecules and molecular weights have nothing to do with this problem. If quantities of acetone and of water are chosen that have equal heats of vaporization, then the sum of the acetone and water flow rates in the vapor stream, expressed in terms of these quantities, will be approximately constant from plate to plate. Departures of the enthalpy curves from the straight lines drawn in Fig. 4-43 will cause variations of approximately 3 per cent in the vapor flow rate expressed in these terms, but this variation is much smaller than the 34 per cent variation in the molal vapor flow rate.

The *method of modified latent heats of vaporization* or the *method of pseudo molecular weights* is therefore often employed [4, 8] as a substantial improvement over the assumption of constant molal overflow. The assumption of constant pseudo molal overflow is still not exact because of the departures from linearity of the saturated liquid and the saturated vapor curves on the enthalpy–composition diagram; but the assumption of constant molal overflow can be in much more serious error, and the additional effort of using the assumption of constant pseudo molal overflow is not very great. Still more accuracy can be achieved by actually making an enthalpy balance, such as Eq. (4-83), employing the actual saturated liquid and saturated vapor curves rather than the straight-line approximations. But this method requires the additional enthalpy data, and it complicates the calculations even more. In some cases, these additional complications are justified when the percentage deviations from linearity of the saturated liquid and vapor curves are large or when small variations in the liquid and vapor rates within the distillation column are important.

With the assumption of constant pseudo molal overflow, the solution of a binary distillation problem using a McCabe–Thiele diagram is quite straightforward. All the material-balance relationships, such as Eqs. (4-9) through (4-12), are unchanged when flow rates are expressed in terms of pseudo moles and compositions are expressed as pseudo mole fractions. First, the equlibrium curve is converted and expressed in terms of pseudo mole fractions in the vapor and liquid phases. Next, the flow rates and compositions of feed and product streams are converted to pseudo moles per unit time and pseudo mole fractions. The McCabe–Thiele diagram construction then proceeds just as in the case in which constant molal overflow is assumed. The q of the feed stream is the pseudo mole fraction liquid, that is, the number of pseudo moles of liquid divided by the number of pseudo moles of liquid plus the number of pseudo moles of vapor in the feed stream. When the feed stream is a subcooled liquid or a superheated vapor, Eqs. (4-15) and (4-16) are interpreted as the heat required to bring a pseudo mole of the feed stream to a saturated vapor condition divided by the heat of vaporization per pseudo mole, and this definition is unambiguous because the heat of vaporization per pseudo mole is constant, independent of composition. The procedure is best illustrated by the following example problems.

EXAMPLE PROBLEM 4-10 Repeat Example Problem 4-5 assuming constant pseudo molal overflow instead of constant molal overflow.

Solution: Since the molal latent heat of varporization of water at its normal boiling point is 1.34 times the molal latent heat of vaporization of acetone at its normal boiling point, the pseudo molecular weight for water will be taken as 18, its actual molecular weight, and the pseudo molecular weight for acetone will be taken as 1.34 times its actual molecular weight of 58.08, that is, 77.9 lb of acetone per pseudo mole of acetone. With this choice, the latent heat of vaporization of water at its normal boiling point and also of acetone at its normal boiling point will be 17,480 Btu per pseudo pound mole.

First, the vapor–liquid equilibrium curve must be converted and expressed in terms of pseudo mole fractions. The vapor–liquid equilibrium curve is given in Fig. 4-19. At $x = 0.55$, the equilibrium value of y is read from that graph to be $y = 0.858$. This point on the equilibrium curve will now be converted to pseudo mole fractions of acetone. First consider the liquid-phase composition, 55 mole per cent acetone. On the basis of one total mole of this liquid phase, the quantity of acetone present is 0.55 mole and the quantity of water present is 0.45 mole. Since 1 mole of water is equivalent to 1 pseudo mole of water, the quantity of water is also 0.45 pseudo moles.

But since the pseudo moleuclar weight of acetone is taken to be 1.34 times the actual molecular weight, the number of pseudo moles of acetone is

$$\frac{0.55}{1.34} = 0.4105 \text{ pseudo mole of acetone}$$

The number of total pseudo moles is

$$0.4105 \text{ pseudo mole of acetone} + 0.45 \text{ pseudo mole of water} = 0.8605 \text{ total pseudo moles}$$

and the acetone pseudo mole fraction is

$$\mathscr{X} = \frac{0.4105}{0.8605} = 0.477$$

Next, the vapor-phase composition is converted to an acetone pseudo mole fraction. Following the same procedure, the vapor-phase mole fraction of 0.858 is converted to a pseudo mole fraction as

$$\mathscr{Y} = \frac{0.858/1.34}{0.142 + 0.858/1.34} = 0.818$$

Therefore, the point on the equilibrium curve ($y = 0.858$, $x = 0.55$) is converted to acetone pseudo mole fractions and becomes the point $\mathscr{Y} = 0.818$, $\mathscr{X} = 0.477$). In a similar manner, a number of other points on the equilibrium curve are converted from acetone mole fractions to acetone pseudo mole fractions, and the entire equilibrium curve is reconstructed, the result being that shown in Fig. 4-44.

Next, the feed and product compositions and flow rates must be converted to the pseudo molal basis. The molal flow rates of acetone and of water in the feed and product streams are given in Table 4-1. The pseudo molal flow rates of water are the same as the actual molal rates, but the pseudo molal flow rates of acetone are obtained by dividing the actual molal rates by 1.34, as shown in Table 4-2. The

Table 4-2

	Pseudo Molal Flow Rate (Pseudo moles/unit time)		
Component	*Feed*	*Distillate*	*Bottom Product*
Acetone	41.05	40.23	0.821
Water	45.0	2.245	42.755
Total	86.05	42.475	43.576

feed, distillate, and bottom product compositions, in terms of acetone pseudo mole fractions, become

$$\mathscr{X}_F = \frac{41.05}{86.05} = 0.477$$

$$\mathscr{X}_D = \frac{40.23}{42.475} = 0.947$$

$$\mathscr{X}_B = \frac{0.821}{43.576} = 0.01883$$

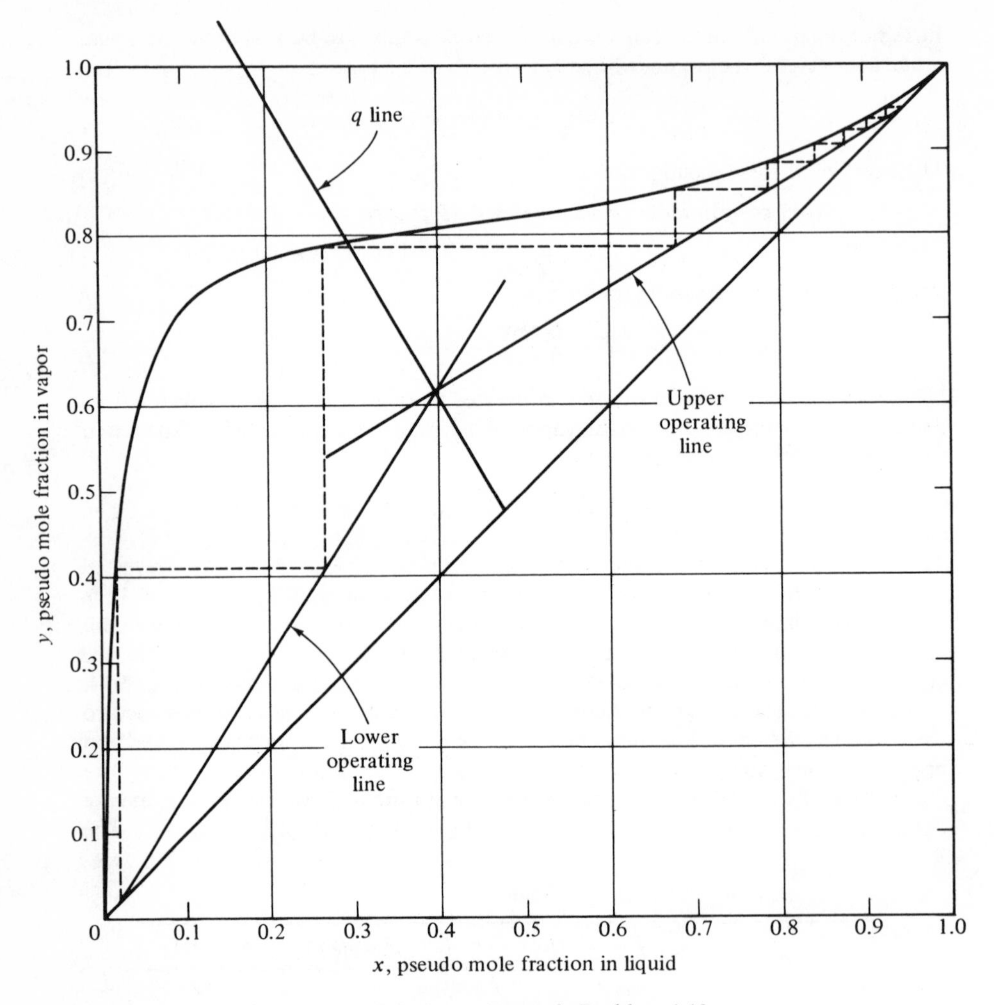

Fig. 4-44. Solution to Example Problem 4-10.

Next it is necessary to calculate the q for the feed stream. In Example Problem 4-5, the feed stream was said to be 60 mole per cent liquid and 40 mole per cent vapor, and therefore q was taken to be 0.6. This approximation was surely as good as the approximation of constant molal overflow adopted in Example Problem 4-5, but in the present context q should be taken to be the pseudo mole fraction liquid in the feed stream. To determine q on this basis, an assumption must be made about the compositions of the vapor and liquid fractions of the feed stream. The overall composition of the feed is 55 mole per cent acetone, which is equivalent to 47.7 pseudo mole per cent acetone. If the feed stream consisted of 0.6 mole of liquid of composition 55 mole per cent acetone plus 0.4 mole of vapor of composition 55 mole per cent acetone, q would be 0.6 whether it was reckoned on a molal basis or on a pseudo molal basis. But in the more likely event that the feed stream consists of a mixture of liquid and vapor of compositions that correspond to equilibrium, q will not equal 0.6 when reckoned on a pseudo molal basis. It will be assumed here that the feed stream is a mixture of liquid and vapor in equilibrium with each other, because this is the most likely case and also will illustrate the ideas involved.

Referring to Fig. 4-19, it is seen that the q line in that diagram intersects the equilibrium curve at the point ($y = 0.838, x = 0.356$). This point represents the compositions of the vapor and liquid fractions of the feed stream, assuming that the vapor and liquid fractions are at equilibrium. The reader may wish to prove the generality that the intersection point of the q line and the equilibrium curve represents the compositions of the vapor and liquid fractions of the feed stream, assuming that they are in equilibrium. Even if the vapor and liquid fractions are not in equilibrium, the point corresponding to their compositions will lie on the q line, as is apparent from Eq. (4-29). Therefore, with the present assumption that the vapor and liquid fractions of the feed stream are in equilibrium, 1 mole of feed contains 0.6 mole of liquid, which is 35.6 mole per cent acetone, and 0.4 mole of vapor, which is 83.8 mole per cent acetone. On the basis of 1 mole of total feed, these quantities are readily converted to pseudo moles, with the result shown in Table 4-3. Therefore,

Table 4-3

	Moles		*Pseudo Moles*	
Component	*Liquid Fraction*	*Vapor Fraction*	*Liquid Fraction*	*Vapor Fraction*
Acetone	0.2135	0.3352	0.1593	0.250
Water	0.3865	0.0648	0.3865	0.0648
Total	0.6000	0.4000	0.5458	0.3148

on the basis of 1 total mole of feed, the liquid fraction of the feed amounts to 0.5458 total pseudo mole of liquid and the vapor fraction amounts to 0.3148 total pseudo mole of vapor. The q of the feed stream, expressed as the pseudo mole fraction liquid in the feed stream, is therefore determined as the pseudo moles of liquid in the feed divided by the total pseudo moles in the feed stream:

$$q = \frac{0.5458}{0.5458 + 0.3148} = 0.635$$

The slope of the q line on the McCabe–Thiele diagram is

$$\frac{-q}{1 - q} = \frac{-0.635}{1 - 0.635} = -1.74$$

The q line can now be located on the McCabe–Thiele diagram as a line with a slope of -1.74 which intersects the 45° diagonal line at a pseudo mole fraction of acetone equal to 0.477. The q line thus located is shown in Fig. 4-44.

The compositions of the liquid and vapor fractions of the feed stream can be readily expressed in terms of acetone pseudo mole fractions by using the values in Table 4-3:

$$\mathscr{X} = \frac{0.1593}{0.5458} = 0.292$$

$$\mathscr{Y} = \frac{0.250}{0.3148} = 0.794$$

Therefore, the point ($\mathscr{Y} = 0.794$, $\mathscr{X} = 0.292$) is a point on the converted equilibrium curve which represents the composition of the vapor and liquid fractions of the feed stream. It is seen in Fig. 4-44 that the q line intersects the equilibrium curve at this point, as it must because the generality of this intersection surely applies whether the diagram is expressed in terms of mole fractions or in terms of pseudo mole fractions.

With a total condenser, the distillate product and the liquid reflux are of the same composition; therefore, the reflux ratio has the same value whether expressed in moles or pseudo moles. With a reflux ratio of 1.5, the liquid-to-vapor flow-rate ratio in the enriching section is given by Eq. (4-18), now interpreted in terms of flow rates expressed as pseudo molal flow rates:

$$\frac{\mathscr{L}}{\mathscr{V}} = \frac{1.5}{2.5} = 0.6$$

Therefore, the upper operating line is located as a line with a slope equal to 0.6 which intersects the 45° diagonal line at an acetone pseudo mole fraction of 0.947. The lower operating line is then drawn through the intersection of the upper operating line and the q line and also through its intersection with the 45° diagonal line at an acetone pseudo mole fraction of 0.01883.

The operating lines and the construction of the steps are shown in Fig. 4-44. Within the accuracy of this graphical plate count, it is seen that an equilibrium reboiler plus seven theoretical plates are required, with the feed stream being fed to the first theoretical plate. This is the same as the result obtained in Example Problem 4-5; therefore, within the limited accuracies of the two plate counts, the result is unchanged.

It is instructive to calculate the liquid and vapor flow rates throughout the column. The liquid rate in the stripping section is given by the product of the reflux ratio and the distillate rate,

$$\mathscr{L} = (1.5)(42.475) = 63.7$$

The vapor rate is then given by Eq. (4-11), written in terms of pseudo molal rates:

$$\mathscr{V} = 63.7 + 42.475 = 106.175$$

Applying Eqs. (4-13) and (4-14) yields

$$\mathscr{L}' = 63.7 + (0.635)(86.05) = 118.3$$

$$\mathscr{V}' = 106.175 - (0.365)(86.05) = 74.73$$

Therefore the pseudo molal vapor rate from the reboiler is 74.73. Referring to Fig. 4-44, this vapor is 41 pseudo mole per cent acetone. Therefore, the actual molal vapor rate from the reboiler is

$$V'_0 = (1.34)(0.41)(74.73) + (0.59)(74.73) = 85.2$$

In Example Problem 4-5 it was concluded that the vapor rate in the entire stripping section of the column would be 100.4. It is seen here that the actual molal vapor rate from the reboiler will be 85.2, with the vapor rate increasing as the acetone mole fraction increases up the column.

EXAMPLE PROBLEM 4-11 Repeat Example Problem 4-7 but making the assumption of constant pseudo molal overflow instead of constant molal overflow.

Solution: The equilibrium curve, the q line, and the distillate and bottom product compositions are the same as they were in Example Problem 4-10. The operating line, corresponding to a tangent pinch, is shown in Fig. 4-45. The slope of the upper operating line is 0.391, and therefore the reflux ratio is

$$\mathscr{R} = \frac{0.391}{1 - 0.391} = 0.642$$

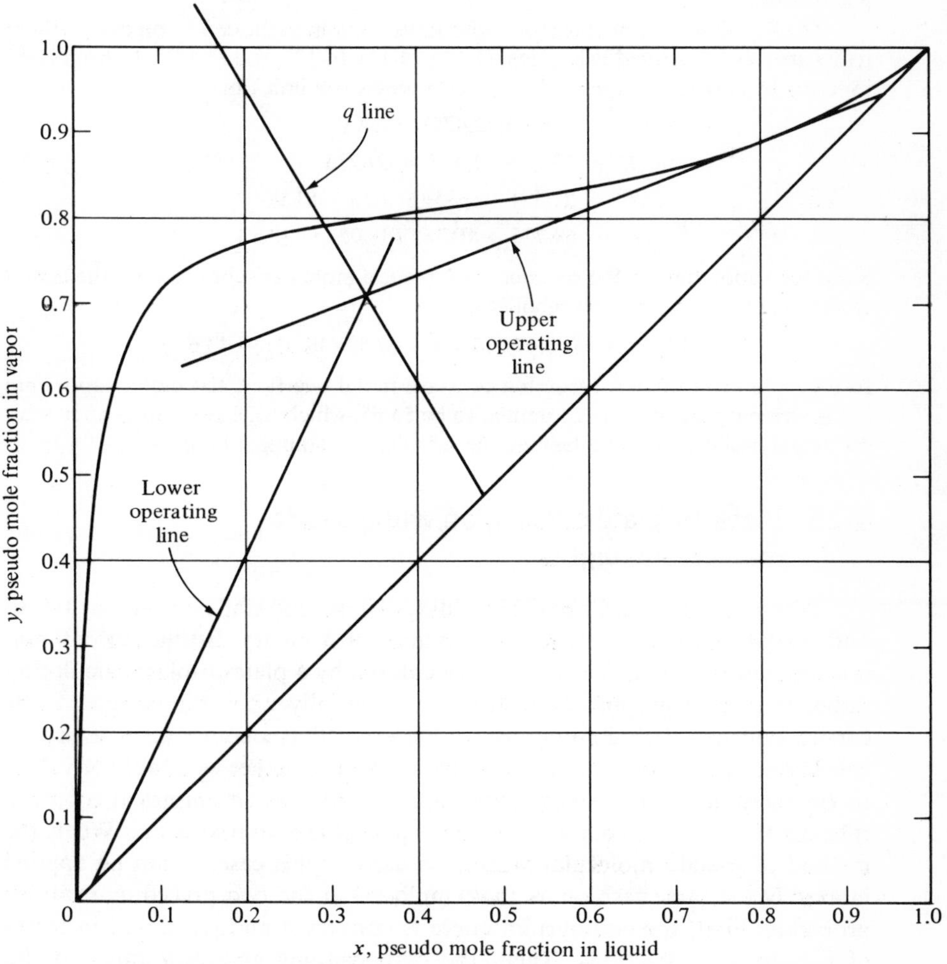

Fig. 4-45. Solution to Example Problem 4-11.

Here the reflux ratio is defined as the liquid-overflow rate in the enriching section of the column divided by the distillate product rate, both rates being expressed in pseudo moles per unit time. More specifically, the reflux ratio is equal to the liquid reflux rate to the *top plate* of the column divided by the distillate product rate; in this case, the rates may both be expressed in pseudo moles per unit time, or they may both be expressed in actual moles per unit time. With a total condenser the liquid reflux to the top plate and the distillate product stream have the same composition, and the ratio of their rates is the same whether the flow rates are expressed as pseudo molal rates or as actual molal rates.

The result obtained here is that the minimum reflux ratio is 0.642. The result obtained in Example Problem 4-7, assuming constant molal overflow throughout the column, was a reflux ratio of 0.704, which is 10 per cent greater than the value computed here.

The liquid and vapor rates throughout the column at the condition of minimum reflux are next computed using Eqs. (4-17), (4-11), (4-13), and (4-14), all formulated in terms of flow rates expressed as pseudo moles per unit time:

$$\mathscr{L} = (0.642)(42.475) = 27.3$$

$$\mathscr{V} = 27.3 + 42.475 = 69.775$$

$$\mathscr{L}' = 27.3 + (0.635)(86.05) = 81.9$$

$$\mathscr{V}' = 69.775 - (0.365)(86.05) = 38.33$$

Since the vapor leaving the reboiler is 41 pseudo mole per cent acetone, the actual molal vapor rate leaving the reboiler is

$$V'_0 = (1.34)(0.41)(38.33) + (0.59)(38.33) = 43.6$$

In Example Problem 4-7, assuming constant molal overflow, the molal vapor rate in the stripping section was computed to be 55.65, which is 28 per cent greater than the actual molal vapor rate leaving the reboiler as computed here.

4-25. Plate-to-plate calculation with pseudo molecular weights

Although the McCabe–Thiele diagram is unparalleled as an aid to understanding the design and performance of a binary distillation column, it is often more convenient to design a column by a plate-to-plate calculation rather than by a graphical construction, especially when the design can be carried out on a digital computer. In the event that a computer is used, the equilibrium curve must be stored in the computer, either as a table of values to be interpolated or, usually more conveniently, as an empirical equation relating the relative volatility to the liquid-phase composition. When the method of pseudo molecular weights is used in this case, it can be applied in exactly the same fashion as that employed in the two preceding example problems. First, the equilibrium curve is converted and expressed in terms of pseudo-mole fractions. Then, the compositions and flow rates of the feed and product streams are likewise converted to the pseudo molal basis.

After converting the q of the feed stream, the plate-to-plate calculation proceeds to accomplish the exact same calculation as that accomplished graphically in Example Problem 4-10.

Alternatively, when the calculation is performed on a digital computer, it might be more convenient to store the equilibrium curve in its normal fashion in terms of mole fractions rather than pseudo mole fractions. In this case, the material balance would be computed in terms of pseudo mole fractions and then the pseudo mole fractions would be converted to real mole fractions to read the equilibrium curve. The result of reading the equilibrium curve would then be converted from mole fractions back to pseudo mole fractions, and the application of the material balance would then proceed in terms of pseudo molal quantities. This procedure will generally require more computer time than the one referred to previously, especially if many column designs are performed with a given equilibrium curve. But the programming time might be reduced by this procedure, and especially in the case of multicomponent distillation problems, the procedure will be found to be more convenient.

4-26. Invariance of the relative volatility

It is useful to realize that for a given liquid composition and equilibrium vapor composition, the relative volatility is the same whether the compositions are expressed as mole fractions or pseudo mole fractions or indeed as weight fractions. This is evident from Eq. (4-1) because the conversion factors for converting from mole fractions to pseudo mole fractions cancel out of the expression leaving the value of α unchanged.

Thus, for a system with a constant relative volatility, the y–x diagram is unchanged by the conversion to pseudo mole fractions. But when α varies with x, the conversion to pseudo mole fractions will alter the y–x diagram because the curve of α versus $\mathscr{X}$ will be different from the curve of α versus x.

4-27. Advantages of the method of modified latent heats

If the molal latent heats of vaporization of the pure components at their boiling points at the column pressure are appreciably different from each other, the assumption of constant molal overflow will be inaccurate. The simple method of using pseudo molecular weights such that the pseudo molal latent heats of vaporization of the two components are equal will correct this error with little additional computational work. Furthermore, as will be seen in Chap. 5, this method is readily extended to the distillation of multicomponent mixtures.

The enthalpy–pseudo mole fraction diagram, such as that shown in

Fig. 4-43, will have saturated liquid and saturated vapor curves which depart somewhat from the horizontal-line approximation of Fig. 4-43. Generally, these departures will be small, and thus the method of modified latent heats will be quite accurate. The departures from the straight-line approximations can become significant when the heat of mixing in the liquid phase is large, or when the difference between the boiling temperatures of the components to be separated is large, the bubble-point curve is nonlinear, and the pseudo molal liquid heat capacities of the components are quite different. In such cases, the pseudo molal flow rates of liquid and vapor within the various sections of the distillation column will not be constant, and a rigorous design of the column would have to consider variations in the flow rates. The vapor and liquid rates could be computed at every plate by use of enthalpy balances such as Eq. (4-83). Alternatively, the graphical construction of Ponchon and Savarit can be employed to accomplish the enthalpy balance graphically. In either event, the necessary data required to define the enthalpy–composition diagram must be available.

The graphical method of Ponchon and Savarit [6, 7, 10] is not applicable to multicomponent distillation problems, and it is somewhat cumbersome to use for binary distillation problems, as is any graphical method which is actually used for design calculations. But like the McCabe–Thiele diagram, the Ponchon–Savarit diagram is quite useful as an aid to understanding and visualizing certain enthalpy effects in distillation systems: for example, the effect of heat losses or additions to the distillation column. The Ponchon–Savarit method will not be discussed here, but it is recommended for future study.

For the majority of the distillation problems normally encountered, most of the error resulting from the assumption of constant molal overflow can be removed by using the method of modified latent heats. Since the method is easy to use, requires only latent heat data for the pure components, and is readily extended to multicomponent distillation problems, this method is recommended for most distillation-column designs. It is advisable, nevertheless, to attempt to assess the maximum departure from linearity of the saturated liquid curve and the saturated vapor curve, expressed as a percentage of the latent heat of vaporization, in order to maintain a cognizance of the percentage variation in the pseudo molal overflow rate for any given system.

ECONOMIC BALANCE

In designing a distillation column to separate a given feed stream into given distillate and bottom product streams, any value of the reflux ratio greater than the minimum reflux ratio may be chosen. For each reflux ratio chosen, there will correspond a different number of theoretical plates required to effect the specified separation. Qualitatively, the curve of the number of theoretical plates required versus the reflux ratio chosen will take the form of the curve shown in Fig. 4-46. As the reflux ratio approaches infinity, the required number of theoretical plates approaches the minimum number of theoretical plates, corresponding to total reflux. As the reflux ratio approaches the minimum reflux ratio, the number of theoretical plates required approaches infinity.

The vapor rates in the stripping and enriching sections vary linearly with the reflux ratio. Combining Eqs. (4-17), (4-11), and (4-14) yields

$$V = (1 + \mathscr{R})D \tag{4-85}$$

$$V' = (1 + \mathscr{R})D - (1 - q)F \tag{4-86}$$

Thus an increase in the reflux ratio corresponds to an increase in the vapor rate within the column, and this requires an increase in the column diameter to keep the vapor velocity below the maximum allowable value. Generally speaking, the maximum allowable vapor velocity in the column is dictated by the requirement that the entrainment of liquid droplets by the upward-flowing vapor stream will be kept acceptably low. Therefore, the cross-sectional area of the column will generally increase approximately linearly with the reflux ratio. But the number of plates required in the column is a decreasing function of the reflux ratio, as shown in Fig. 4-46. Therefore, the cost of the column will pass through a minimum value as the reflux ratio

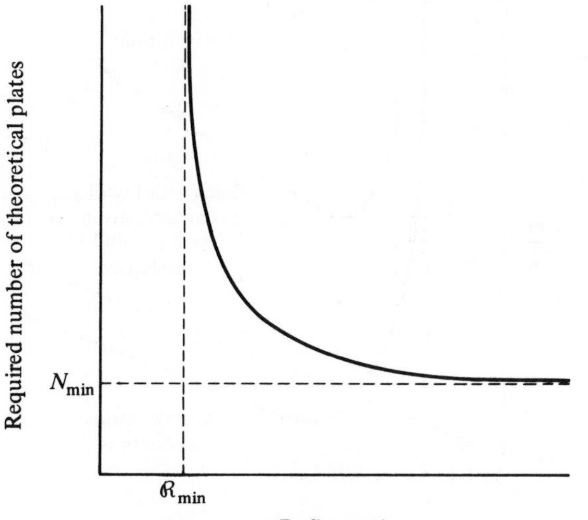

Fig. 4-46. Distillation design curve.

is varied. The cost of the reboiler and the condenser and the cost of energy supplied to the reboiler and the coolant supplied to the condenser will increase approximately directly as the vapor rate increases; therefore, these costs will increase with the reflux ratio.

Figure 4-47 shows a qualitative plot of annual costs versus the reflux ratio. The cost of energy and cooling plus the amortization on the cost of the reboiler and condenser are shown as a rising function of the reflux ratio. The amortization on the cost of the column approaches infinity as the reflux ratio approaches its minimum value, and this cost shows a minimum with respect to the reflux ratio. The total cost, which is the sum of the two curves, also passes through a minimum value at a reflux ratio smaller than the reflux ratio at which the column cost passes through a minimum. The minimum in the total cost curve corresponds to the optimum design for the given distillation system.

An important contribution to the economic balance curves shown in Fig. 4-47, and indeed the only contribution to which cascade theory is relevant, is the curve of the number of plates required versus the reflux ratio, which is shown in Fig. 4-46. It is likewise obvious that the minimum reflux ratio, $\mathscr{R}_{\min}$, and the minimum number of theoretical plates, $N_{\min}$, dictate, in an approximate way, the cost of a distillation system. The entire curve in Fig. 4-46 is scaled by the parameters $\mathscr{R}_{\min}$ and $N_{\min}$. Generally speaking, the number of theoretical plates required at the optimum point will often be of the order of twice the minimum number of theoretical plates, and the

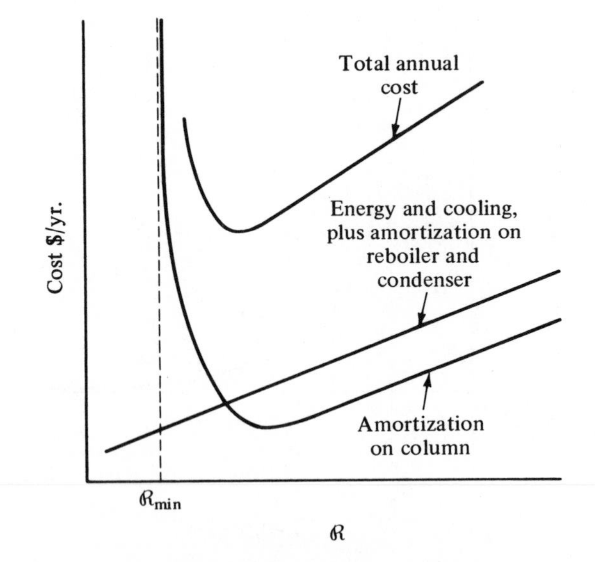

Fig. 4-47. Economic balance curves.

reflux ratio at the optimum point will be of the order of 1.1 to 1.5 times the minimum reflux ratio.

Indeed, Gilliland [2, 9] has shown that the entire curve of the number of theoretical plates required versus the reflux ratio is dictated, to a good approximation, by the values of $\mathscr{R}_{\min}$ and $N_{\min}$. Figure 4-48 shows the Gilliland correlation, which relates the number of stages required to the reflux ratio, the minimum reflux ratio, and the minimum number of stages. Here $\mathscr{N}$ is the total number of equilibrium *stages* required for the separation. If a partial reboiler and a total condenser are used, $\mathscr{N} = N + 1$. If a partial condenser is employed which effects an additional stage of separation, $\mathscr{N} = N + 2$. The curve shown in Fig. 4-48 has been found to be a reasonable approximation to the solution to a wide variety of distillation systems. The single curve is only an approximation; a slightly lower curve would better describe distillation systems with q equal to unity, and a slightly higher curve would better describe systems with q equal to zero. But the single curve is a useful approximation for making preliminary economic-balance calculations. Furthermore, if a single plate-to-plate calculation has been made, it is often useful to plot a point corresponding to that calculation on the generalized graph and to draw through that point a curve of shape similar

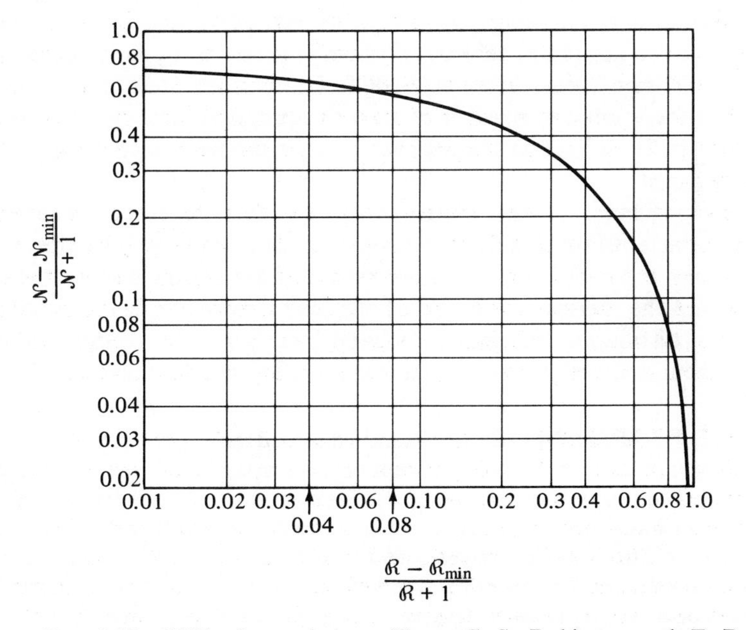

Fig. 4-48. Gilliland correlation. (From C. S. Robinson and E. R. Gillaland, *Elements of Fractional Distillation*, 4th ed., New York, McGraw-Hill, 1950.)

to the curve shown in the generalized correlation. This procedure will usually result in a correlation of good accuracy for that system. The generalized correlation is especially useful as an approximate design method for multicomponent distillation problems.

The fact that $\mathscr{R}_{\min}$ and $N_{\min}$ approximately determine the entire curve of N versus $\mathscr{R}$, as represented by the generalized correlation of Fig. 4-48, emphasizes the role played by these minimum quantities in determining the cost of a distillation system. Accordingly, it is useful to realize that, while $N_{\min}$ is in general a function of both the relative volatility and the degree of separation desired, $\mathscr{R}_{\min}$ is a function of the relative volatility and the nature of the feed stream but is approximately independent of the degree of separation desired. The effects of the relative volatility of the system and of the degree of separation desired, as measured by $\mathscr{S}$, upon the minimum number of stages required are readily visualized from Eq. (4-43). Furthermore, it should be noted that the composition and the thermal condition of the feed stream have no effect upon the system at total reflux and therefore do not affect $N_{\min}$. On the other hand, the minimum reflux ratio is determined mainly by the nature of the vapor–liquid equilibrium curve and by the composition and the thermal condition of the feed stream. The effect of the distillate and bottom product stream purities is small, the effect essentially vanishing as the product streams are made quite pure. This is readily visualized in Fig. 4-16. For a given vapor–liquid equilibrium curve and for given values of q and z_F, the minimum reflux ratio hardly varies as x_D is increased from 0.99 to 0.999 and then further to 0.9999. In contrast, it is clear from Eq. (4-43) that the minimum number of stages required at total reflux continues to increase without limit as the product purities are pushed closer and closer to 100 per cent.

The significance of these generalizations is that the energy requirement and the diameter of the distillation column are dictated largely by the relative volatility and the molal heat of vaporization of the system and by the composition and the thermal condition of the feed stream, while the number of plates in a distillation column is dictated largely by the relative volatility of the system and the purity requirements on the product streams.

EXAMPLE PROBLEM 4-12 One hundred pound moles per hour of a saturated liquid mixture containing 35 mole per cent benzene and 65 mole per cent toluene is to be distilled continuously into a liquid distillate product which is 99 mole per cent benzene and a liquid bottom product which is 1 mole per cent benzene. The column will operate at atmospheric pressure, and it will be fitted with a partial reboiler and a total condenser. The overall column efficiency is estimated to be 80 per cent, but the reboiler will be assumed to give an equilibrium stage of separation.

(a) Determine the economic optimum design for this distillation column, based upon the following approximations:

1. The column diameter will be chosen so that the superficial vapor velocity (based upon the column cross-sectional area) will not exceed 1.0 ft/sec.

2. The installed cost of the column is estimated to be

$$\text{cost in dollars/actual plate} = 2.1(\text{diameter in inches})^{1.57},$$

3. The installed cost of the reboiler plus the condenser is estimated to be

$$\text{cost in dollars} = 1.7(Q_R)^{0.6}$$

where Q_R is the reboiler heat load in Btu/hr.

4. The cost of the heat supplied to the reboiler plus the cooling in the condenser is estimated to be $0.70 per million Btu supplied to the reboiler.

5. An operating year of 8500 hr is assumed.

6. The total capital investment is estimated to be 1.5 times the installed cost of the column, reboiler, and condenser. This will include pumps, piping, and instrumentation.

7. The amortization rate on invested capital is taken as 20 per cent per year to account for depreciation, interest on invested capital, and maintenance.

8. The total annual cost will be taken as the amortization on the capital investment plus the cost of heat and cooling (the labor cost is fixed, independent of the column design).

At 1 atm the latent heat of vaporization is 14,400 Btu/lb mole for toluene and 13,260 Btu/lb mole for benzene. Use the method of pseudo molecular weights.

(b) Use the Gilliland correlation to determine NAP as a function of $\mathscr{R}$ and compare this approximation with the relationship developed as part of the solution to part (a).

Solution

(a) The molal vapor rate at the top of the column will be approximately 9 per cent higher than that at the bottom because benzene has the smaller molal latent heat of vaporization. On the other hand, the absolute temperature at the top of the column will be approximately 9 per cent lower than that at the bottom because benzene has the lower boiling point. Using the ideal-gas law, these two effects cancel each other almost quantitatively, and the *volumetric* flow rate of vapor is essentially constant throughout the column, assuming that pressure is constant at 1 atm. Choosing a pseudo mole of toluene to be an actual mole and a pseudo mole of benzene to be 1.09 actual moles, the volumetric flow rate of vapor can therefore be related to the pseudo molal vapor rate by using the vapor density of pure toluene at its boiling point. Assuming the ideal-gas law to be an adequate approximation, the molal volume of toluene at its boiling point is

$$359\left(\frac{384}{273}\right) = 505 \text{ ft}^3\text{/lb mole}$$

The column diameter in inches, d, is therefore given by

$$\frac{(\mathscr{V} \text{ pseudo lb moles/hr})(505 \text{ ft}^3\text{/psuedo lb mole})}{(3600 \text{ sec/hr})\,[(\pi/4)(d/12)^2 \text{ ft}^2]} = 1.0 \text{ ft/sec}$$

where $\mathscr{V}$ is the pseudo molal vapor rate in the column. Solving for the column diameter, the column cost per plate is easily related to $\mathscr{V}$.

The reboiler heat duty is given by

$$Q_R(\text{Btu/hr}) = \left(\mathscr{V}\frac{\text{pseudo lb moles}}{\text{hr}}\right)\left(\frac{14{,}400 \text{ Btu}}{\text{pseudo lb mole}}\right)$$

Table 4-4

ECONOMIC BALANCE ON BENZENE–TOLUENE DISTILLATION COLUMN

Reflux Ratio	*Actual Plates*	*Feed Plate*	*Column Diameter*	*Cost of BLR + CDSR*	*Cost of Column*	*Cost of Heat + Cool*	*Total Cost*
2.1000	38.5089	22.5000	50.4520	2511.32	11439.84	8480.15	22431.31
2.1500	36.4260	21.2500	50.7063	2526.51	10906.83	8565.83	21999.17
2.2000	33.6790	20.0000	51.2109	2556.72	10242.33	8737.19	21536.23
2.2500	32.4766	18.7500	51.4614	2571.73	9952.59	8822.87	21347.19
2.3000	30.8212	17.5000	51.9588	2601.59	9589.01	8994.23	21184.82
2.3500	29.5553	16.2500	52.4514	2631.22	9332.42	9165.59	21129.22
2.4000	28.7983	16.2500	52.6960	2645.95	9160.04	9251.27	21057.25
2.4500	27.8995	16.2500	53.1818	2675.24	9002.94	9422.63	21100.81
2.5000	27.3921	15.0000	53.4230	2689.81	8902.24	9508.30	21100.36
2.5500	26.5266	15.0000	53.9023	2718.79	8742.68	9679.67	21141.14
2.6000	25.9010	15.0000	54.3773	2747.57	8654.90	9851.03	21253.50
2.6500	25.5892	13.7500	54.6133	2761.88	8609.04	9936.71	21307.63
2.7000	24.8527	13.7500	55.0822	2790.36	8474.25	10108.07	21372.68
2.7500	24.6124	13.7500	55.3151	2804.53	8448.09	10193.75	21446.37
2.8000	24.1190	13.7500	55.7781	2832.72	8387.80	10365.11	21585.63

Therefore, the cost of the reboiler plus the condenser is simply related to $\mathscr{V}$. Multiplying these costs by 1.5 and then taking 20 per cent of them converts them to annual amortization costs in dollars per year. The result is

$$\text{cost, dollars/yr} = (0.2)(1.5)(2.1)\left[\frac{(4)(144)(505)\mathscr{V}}{3600\pi}\right]^{0.785}(\text{NAP})$$

for the column.

$$\text{cost, dollars/yr} = (0.2)(1.5)(1.7)(14{,}400\mathscr{V})^{0.6}$$

for the reboiler plus the condenser.

The operating costs for heating and cooling are

$$\left(14{,}400\mathscr{V}\,\frac{\text{Btu}}{\text{hr}}\right)\left(8500\,\frac{\text{hr}}{\text{yr}}\right)(\$0.70 \times 10^{-6}\text{/Btu})$$

$$= (14{,}400)(8500)(0.7 \times 10^{-6})\mathscr{V}\ \text{dollars/yr}$$

A digital-computer program was written to solve the distillation-column design by a plate-to-plate calculation, assuming constant pseudo molal overflow as outlined in this chapter. The equilibrium curve of Fig. 4-13 was fitted by a broken-line relationship between α and x_b. For a given reflux ratio the number of theoretical plates was divided by 0.8 to yield NAP. The cost formulas above were then employed to determine the annual cost. The details will not be given here, but the results are given in Table 4-4 and Fig. 4-49.

Since the number of actual plates must be an integer, the optimum design is either

$$\mathscr{R} = 2.385, \qquad \text{NAP} = 29$$

or

$$\mathscr{R} = 2.445, \qquad \text{NAP} = 28$$

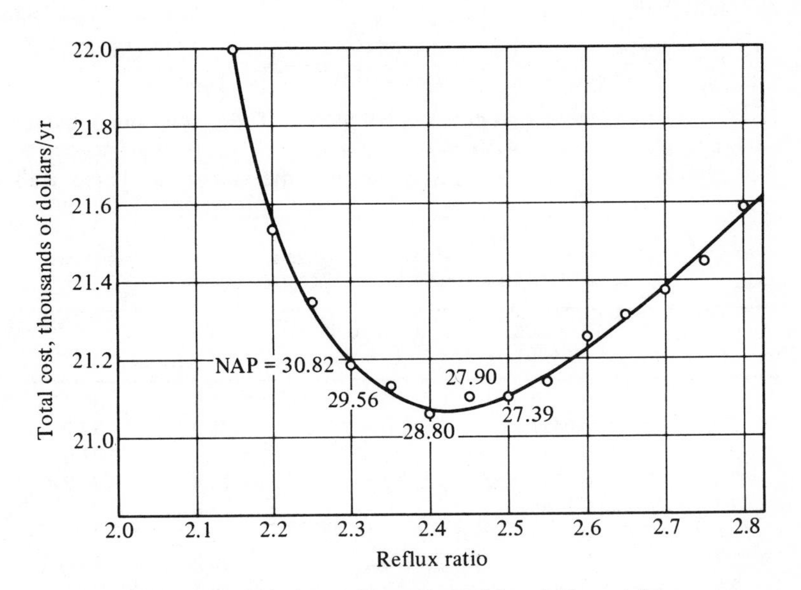

Fig. 4-49. Solution to Example Problem 4-12, part (a).

For either design, the annual cost is between \$21,060 and \$21,080.

(b) For total reflux, the Fenske equation, Eq. (4-43), will be used with $\bar{\alpha} = 2.41$, as determined in the solution to Example Problem 4-4:

$$\mathcal{N}_{\min} = \frac{\ln\left(\dfrac{0.99/0.01}{0.01/0.99}\right)}{\ln 2.41} = 10.45$$

Note that actual mole fractions of the products can be used in the Fenske equation; had pseudo mole fractions been used, the answer would have been the same.

For the minimum reflux ratio, pseudo mole fractions should be used. Converting the distillate product composition

$$\mathscr{X}_D = \frac{99/1.09}{99/1.09 + 1} = 0.98911$$

Reading the equilibrium curve in Fig. 4-13 at $x = 0.35$ gives $y = 0.568$. Converting this point on the equilibrium curve to pseudo mole fractions gives

$$\mathscr{Y} = \frac{0.568/1.09}{0.568/1.09 + 0.432} = 0.546$$

$$\mathscr{X} = \frac{0.35/1.09}{0.35/1.09 + 0.65} = 0.3305$$

Therefore, the point ($\mathscr{Y} = 0.546$, $\mathscr{X} = 0.3305$) represents the intersection of the q line with the equilibrium curve on the pseudo-mole-fraction basis.

At minimum reflux, the slope of the upper operating line is

$$\frac{\mathscr{L}}{\mathscr{V}} = \frac{0.98911 - 0.546}{0.98911 - 0.3305} = 0.673$$

The minimum reflux ratio is

$$\mathscr{R}_{\min} = \frac{0.673}{1 - 0.673} = 2.06$$

Using Fig. 4-48, the design curve is computed. The results are tabulated in Table 4-5 and plotted in Fig. 4-50.

Figure 4-50 shows the comparison between the values calculated with the

Table 4-5
RESULTS FROM THE GILLILAND CORRELATION

$\mathscr{R}$	$\dfrac{\mathscr{R} - 2.06}{\mathscr{R} + 1}$	$\dfrac{\mathcal{N} - 10.45}{\mathcal{N} + 1}$	$\mathcal{N}$	NAP
2.1	0.0129	0.71	38.5	46.9
2.2	0.0437	0.64	30.8	37.2
2.3	0.0727	0.58	26.3	31.6
2.4	0.10	0.55	24.4	29.2
2.5	0.126	0.52	22.8	27.2
2.6	0.15	0.48	21	25
2.7	0.173	0.46	20.2	24
2.8	0.195	0.44	19.5	23.1

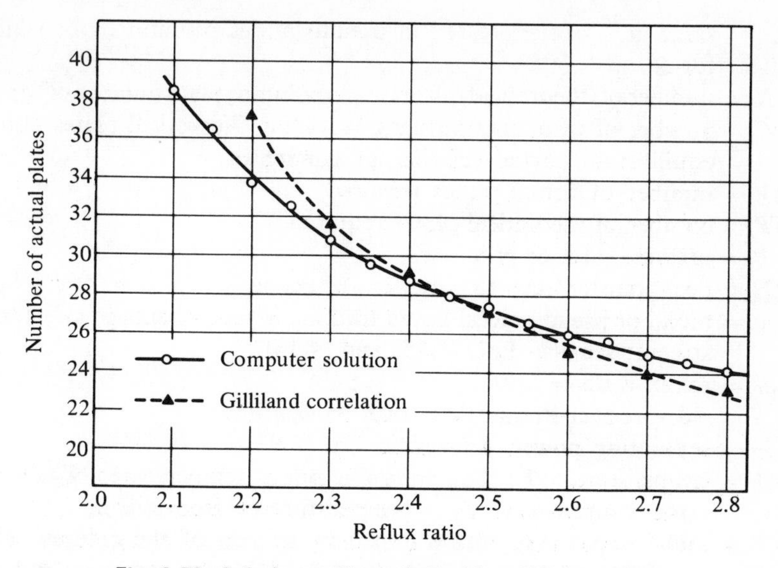

Fig. 4-50. Solution to Example Problem 4-12, part (b).

Gilliland correlation and those obtained from the plate-to-plate calculation using the computer. It is seen that the Gilliland correlation does provide a useful approximation to the design curve.

NOMENCLATURE

B = bottom product flow rate, moles/time

c_{pL} = molal heat capacity of liquid feed, Btu/mole °F

c_{pV} = molal heat capacity of vapor feed, Btu/mole °F

D = distillate product flow rate, moles/time

d = column diameter, inches

$\mathscr{E}_{col}$ = overall column efficiency; see Eq. (4-81)

$\mathscr{E}_{plt}$ = plate efficiency or Murphree vapor efficiency; see Eq. (4-82)

F = molal flow rate of feed "F", moles/time

G = molal flow rate of feed "G", moles/time

H = molal enthalpy of vapor, Btu/mole; H_n refers to vapor stream *leaving* nth plate

h = molal enthalpy of liquid, Btu/mole; h_n refers to liquid stream leaving nth plate

L = molal liquid overflow rate within enriching section of the column, moles/time; L' and L'' are similarly defined for stripping and intermediate sections of the column, respectively; L_n refers to liquid stream leaving nth plate

$\mathscr{L}$ = same as L but expressed in pseudo moles per unit time; similarly for $\mathscr{L}'$ and $\mathscr{L}''$

N = number of (theoretical) plates in the column; plate number of top plate

$\mathscr{N}$ = number of equilibrium stages, including theoretical plates and also equilibrium partial reboiler or condenser

NAP = number of actual plates required

NTP = number of theoretical plates required

P = pressure, atm or psia

Q_R = heat-transfer rate to reboiler, Btu/time

q = molal or pseudo molal liquid fraction in feed stream or side product stream; see also Eqs. (4-15) and (4-16)

$\mathscr{R}$ = reflux ratio $= L/D$

S = side-product-stream flow rate, moles/time

$\mathscr{S}$ = separating power, defined in Eq. (4-42)

T = temperature, °F; T_{BP} = bubble-point temperature; T_{DP} = dew-point temperature; T_F = temperature of feed stream

V = molal vapor rate within enriching section of the column, moles/time; V' and V'' are similarly defined for the stripping and intermediate sections of the column, respectively; V_n refers to vapor stream *leaving* nth plate

$\mathscr{V}$ = same as V but expressed in pseudo moles per unit time; similarly for $\mathscr{V}'$ and $\mathscr{V}''$

x = liquid-phase mole fraction (usually of more volatile component); x_n refers to liquid *leaving* nth plate

$\mathscr{X}$ = liquid-phase pseudo mole fraction.

y = vapor-phase mole fraction (usually of more volatile component); y_n refers to vapor *leaving* nth plate

y_n^* = value of y in equilibrium with x_n

$\mathscr{Y}$ = vapor-phase pseudo mole fraction

z = mole fraction in feed or side product stream; if the stream is a mixture of liquid and vapor, z represents the overall mole fraction in the mixture

$\mathscr{Z}$ = same as z but expressed as a pseudo mole fraction

GREEK LETTERS

α = relative volatility of more volatile component with respect to less volatile component; $\alpha_{b,t}$ = relative volatility of benzene with respect to toluene; see Eqs. (4-1) through (4-6); $\bar{\alpha}$ = average value of α

ΔH = molal latent heat of vaporization, Btu/mole

SUBSCRIPTS

B = bottom product stream

b = benzene

D = distillate product stream
e = ethanol
F = feed stream "F"
f = plate f, the plate to which feed stream "F" is added
G = feed stream "G"
g = plate g, the plate to which feed stream "G" is added
i = point of intersection of two lines
min = minimum value
mix = mixture of two streams
m = plate m, an arbitrary plate in the stripping section
N = plate N, the top plate
n = plate n, an arbitrary plate in the enriching section
na = nitric acid
p = plate p, an arbitrary plate in the intermediate section
R = reboiler
S = side product stream
s = plate s, the plate from which the side product is withdrawn
t = toluene
0 = vapor entering bottom plate
1 = plate 1, the bottom plate
2 = plate 2, the second plate from the bottom

REFERENCES

1. M. R. Fenske, *Ind. Eng. Chem.*, **24**, 482 (1932).
2. E. R. Gilliland, *Ind. Eng. Chem.*, **32**, 1220 (1940).
3. W. K. Lewis, *Ind. Eng. Chem.*, **14**, 492 (1922).
4. W. L. McCabe and E. W. Thiele, *Ind. Eng. Chem.*, **17**, 605 (1925).
5. E. V. Murphree, *Ind. Eng. Chem.*, **17**, 747 (1925).
6. M. Ponchon, *Tech. Moderne*, **13**, 20 (1921).
7. C. S. Robinson and E. R. Gilliland, *Elements of Fractional Distillation*, 4th ed., McGraw-Hill, Inc.: New York, 1950, pp. 146ff.
8. *Ibid.*, pp. 158ff.
9. *Ibid.*, pp. 347ff.
10. R. Savarit, *Arts et Métiers*, 65 (1922).

STUDY PROBLEMS

1. A saturated liquid mixture of benzene and toluene containing 35 mole per cent benzene is to be distilled continuously in a fractional-distillation column operat-

ing at a pressure of 1 atm. The liquid distillate product is to be benzene at a purity of 95 mole per cent, and the liquid bottom product is to be toluene at a purity of 95 mole per cent. The column is to be equipped with a total condenser and a partial reboiler and is to be operated with a reflux ratio of 2.25.

How many theoretical plates must the column contain to accomplish this separation? Where should the feed plate be located (for minimum total number of plates at the stated reflux ratio)?

2. With reference to Problem 1:
   (a) What is the minimum allowable reflux ratio for the stated separation? On the basis of 100 moles of feed, what is the vapor rate, V', leaving the reboiler at this reflux ratio?
   (b) Repeat part (a) for a partially vaporized feed stream which is 50 mole per cent vapor and 50 mole per cent liquid.

3. (a) What is the minimum number of theoretical stages, at a reflux ratio approaching infinity, needed to accomplish the separation of Problem 1? Determine this number *both* graphically and analytically.
   (b) Repeat part (a) for a feed that is 50 per cent vaporized.

4. Write a computer program to solve benzene–toluene distillation problems such as Problem 1. Assume that the column has one feed stream and two liquid-product streams of specified purities and that it has a total condenser and a partial reboiler. Arrange the program to read in the feed composition, the q value of the feed, the specified distillate and bottom product compositions, and the reflux ratio. Assume that the relative volatility is constant at a value of 2.50, and assume constant molal overflow. Arrange the printout to include the reboiler vapor rate per mole of feed, the number of theoretical plates required, the feed-plate location, and the distillate and bottom product rates per mole of feed.

   Run the program to solve Problem 1. Also run the program to solve the column design of Example Problem 4-12 with $\mathscr{R} = 2.385$, and check the number of plates reported there.

5. Using Fig. 4-3, determine the relative volatility of benzene with respect to toluene at several liquid-phase compositions. Plot $\alpha_{b,t}$ versus x_b and obtain a satisfacotry broken-line representation of the curve. Repeat Problem 4 with this representation of the vapor–liquid equilibrium instead of the constant value of α assumed there.

6. Repeat Problem 5 with the additional modification of assuming constant pseudo molal overflow instead of constant molal overflow.

7. Write a computer program to solve Example Problem 4-12, part (a), and use it to check the solution reported.

8. Two saturated liquid benzene–toluene streams are to be distilled into three liquid product streams, as shown in the table on the following page.

   This separation is to be accomplished in a single fractional-distillation column with a side product stream, operating at atmospheric pressure, equipped with a total condenser and a partial reboiler.
   (a) If the vapor rate from the reboiler is 85 lb moles/hr, how many theoretical

Stream	Flow Rate, lb moles/hr	Benzene Mole Fraction
Rich feed	100	0.85
Lean feed	100	0.15
Distillate product	50	0.95
Intermediate product	100	0.50
Bottom product	50	0.05

plates are required for this separation? Specify the feed-plate locations and the side-product withdrawal point.

(b) What is the minimum possible vapor rate required for the separation?

(c) Alternatively, it has been proposed that the intermediate product stream be obtained by simply blending appropriate portions of the two feed streams, the remainder of the two feed streams being sent to the column. For this proposal, repeat parts (a) and (b), and discuss the results.

Use the McCabe–Thiele diagram, not a computer, to solve this problem.

Multicomponent Distillation 5

Most of the separation problems facing the chemical process industries involve multicomponent mixtures to be separated. When such mixtures are to be separated by distillation into essentially pure components, it is necessary to employ a distillation system consisting of several fractional-distillation columns. The first column will involve the distillation of the multicomponent mixtures, while the last columns will generally involve the distillation of essentially binary mixtures.

For example, a mixture of benzene, toluene, and xylene is produced from the coking of coal. If this mixture is to be separated into essentially pure benzene, pure toluene, and pure xylene by continuous distillation, two distillation columns are required. The first column could be used to distill the mixture of benzene, toluene, and xylene into a distillate product which is essentially pure benzene and a bottom product which contains almost all the toluene and xylene. This bottom product from the first column would then be the feed stream to a second column which would produce a distillate product of almost pure toluene and a bottom product of relatively pure xylene. If the distillation in the first column is effective in keeping benzene out of the bottom product, then the feed stream to the second column will contain only a trace of benzene; therefore, the second column may be analyzed as a binary distillation of toluene from xylene. But the first column involves a multicomponent separation because the feed to that column would generally contain appreciable quantities of benzene, toluene, and xylene. In the first column, benzene and toluene would be considered the *key components*, because the objective of the separation would be to minimize the benzene content of the bottom product and the toluene content of the distillate product. But the presence of the xylene will have an effect upon the performance of the column, although the objective of the column is essentially the separation of the benzene from the toluene.

The behavior of a multicomponent distillation column differs in a number of respects from that of a binary distillation column. Furthermore, the presence of the additional components complicates the column design considerably. Even the problem of representing the vapor–liquid equilibrium relationships is very much complicated by the presence of the additional components in the system. Generally speaking, graphical representations such as the McCabe–Thiele diagram are not useful in analyzing multicomponent distillation problems, and the use of a digital computer is almost a necessity.

MULTICOMPONENT VAPOR–LIQUID EQUILIBRIUM

The isobaric vapor–liquid equilibrium relationships in a binary system can be adequately determined by relatively few experimental measurements and can be simply represented by graphs such as those shown in Figs. 4-2 and 4-3. Graphical representations such as these can present the equilibrium relationships over the entire range of compositions in a binary system. As shown in Chap. 4, the essential elements of a binary distillation-column design can be accomplished using the y–x diagram, which represents the vapor–liquid equilibrium relationship with a single curve. This is not possible when dealing with a multicomponent system. Even for a ternary system, the mole fractions of two components must be specified to completely describe the composition of a phase; therefore, one would need a four-dimensional graph to relate the vapor-phase composition to the liquid-phase composition. The representation of experimental vapor–liquid equilibrium data is therefore enormously complicated, and the number of experimental measurements required to determine the equilibrium relationship for even a ternary system is immense. It is usual, therefore, to seek a mathematical description of the equilibrium relationship in a multicomponent system rather than to simply present experimental results in a graphical or tabular fashion. In this connection, the thermodynamic theory of vapor–liquid equilibrium is enormously helpful, and without it the analysis of multicomponent distillation systems would probably have made no progress at all. The thermodynamic theory of vapor–liquid equilibrium in nonideal systems

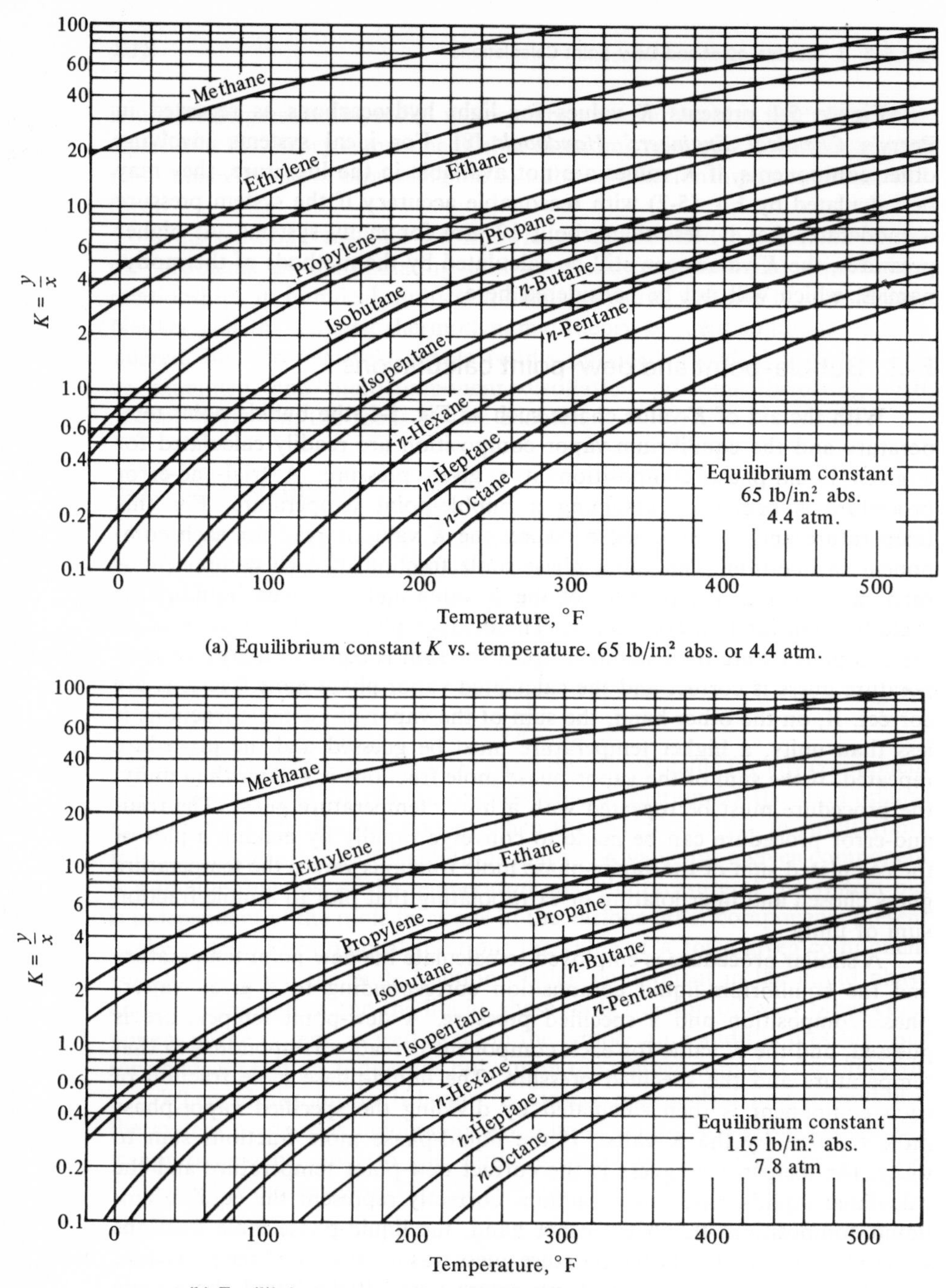

(a) Equilibrium constant K vs. temperature. 65 lb/in.² abs. or 4.4 atm.

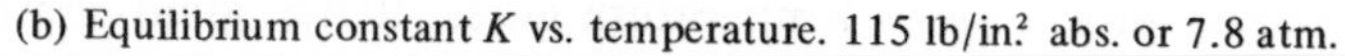

(b) Equilibrium constant K vs. temperature. 115 lb/in.² abs. or 7.8 atm.

Fig. 5-1. K values for light hydrocarbons. (Reproduced from John H. Perry, ed., *Chemical Engineers Handbook*, 3rd ed. New York, McGraw-Hill, Inc., 1950. Used with permission of McGraw-Hill, Inc.)

EXAMPLE PROBLEM 5-1 Calculate the bubble-point temperature and the equilibrium-vapor composition at a pressure of 4.4 atm for a liquid-phase mixture of isobutane, *n*-butane, isopentane, and *n*-pentane with liquid-phase mole fractions of 0.2, 0.3, 0.15, and 0.35, respectively.

Solution: The K values for these components are plotted in Fig. 5-1. In that figure it can be seen that the K values for *n*-butane and isopentane bracket the value of unity at approximately 140°F. Therefore, as a first guess at the bubble-point temperature, 140°F will be chosen. At this value and at a pressure of 4.4 atm, the K values for the four components are read from Fig. 5-1. Multiplying these K values by the specified x values results in the vapor-phase mole fractions shown in Table 5-1. The computed values of y sum to 1.0368, and therefore the actual bubble point is somewhat less than 140°F. From the slope of the K-value curve for *n*-butane it is seen that K changes 4 per cent in approximately 4°F. If the K values for all of the components changed exactly the same percentage for a given change in temperature, the correct bubble-point temperature would be the temperature at which the K values were all 3.7 per cent lower than the values at 140°F. The correct vapor-phase composition for that point could be obtained by simply dividing each of the tabulated values of y by 1.0368. The resulting normalized values are shown in the last column of Table 5-1.

Table 5-1

Component	x	*K at 140°F*	$y = Kx$	$y/1.0368$
Isobutane	0.2	1.82	0.364	0.351
n-Butane	0.3	1.35	0.405	0.3907
Isopentane	0.15	0.63	0.0945	0.0911
n-Pentane	0.35	0.495	0.1733	0.1672
			1.0368	1.0000

As a second temperature guess, 135°F will be chosen. The K values read from Fig. 5-1 and the computed values of y are shown in Table 5-2. This time the computed values of y sum to 0.9855. When each of the values of y is divided by 0.9855, the resulting normalized values, shown in the last column of Table 5-2, show reasonable agreement with the normalized values obtained at 140°F. The differences be-

Table 5-2

Component	x	*K at 135°F*	$y = Kx$	$y/0.9855$
Isobutane	0.2	1.7	0.34	0.345
n-Butane	0.3	1.32	0.396	0.402
Isopentane	0.15	0.59	0.0885	0.0898
n-Pentane	0.35	0.46	0.161	0.1632
			0.9855	1.0000

tween the two sets of values are due in part to errors in reading the K values and also in part to the fact that the K values for the different components do not change exactly the same percentage with a 5°F change in temperature. Within the accuracy attainable with the K values in Fig. 5-1, the solution to this problem can be represented by a bubble-point temperature of 137°F and a vapor-phase composition represented by the normalized values of y shown in the last column in Table 5-2.

EXAMPLE PROBLEM 5-2 Calculate the dew-point temperature and the equilibrium-liquid composition at a pressure of 4.4 atm for a vapor-phase mixture of isobutane, n-butane, isopentane, and n-pentane containing 25 mole per cent of each of these components.

Solution: As a first guess at the dew-point temperature, 140°F will be chosen. Dividing the specified values of y by the K values at 140°F yields the x values. The calculations are summarized in Table 5-3. The computed values of x sum to 1.2247, and therefore the true dew-point temperature is greater than 140°F.

Table 5-3

Component	y	*K at 140°F*	$x = y/K$	$x/1.2247$
Isobutane	0.25	1.82	0.1374	0.1123
n-Butane	0.25	1.35	0.1853	0.1514
Isopentane	0.25	0.63	0.397	0.324
n-Pentane	0.25	0.495	0.505	0.4123
			1.2247	1.0000

As a second guess, 160°F will be chosen. The K values read at this temperature and the computed values of x are summarized in Table 5-4. This time the x values sum to 0.9402, and thus it is seen that the dew-point temperature is somewhat below 160°F. The normalized values of x shown in the last column of Table 5-4 are not too different from the normalized x values obtained at 140°F. Therefore, as an approximation, one could take the equilibrium-liquid composition to be that represented by the normalized x values in Table 5-4. The dew-point temperature is approximately 156°F, interpolating between the two calculations presented above. The accuracy

Table 5-4

Component	y	*K at 160°F*	$x = y/K$	$x/0.9402$
Isobutane	0.25	2.22	0.1127	0.1199
n-Butane	0.25	1.7	0.1472	0.1567
Isopentane	0.25	0.83	0.3013	0.3204
n-Pentane	0.25	0.66	0.379	0.403
			0.9402	1.0000

of this approximation could be checked by repeating the calculation with a temperature guess of 156°F, but this will not be done here. The calculations performed should suffice for illustrative purposes.

5-3. Relative volatility

Referring to the bubble-point and dew-point calculations illustrated in Example Problems 5-1 and 5-2, even if the temperature guess is incorrect, the normalized values of y or x will be approximately correct if the K values for all the components change by approximately the same percentage for a given change in temperature. This would correspond to approximately constant relative volatilities among the various components.

The relative volatility of component j with respect to component k is defined as

$$\alpha_{j,k} \equiv \left[\frac{y_j/x_j}{y_k/x_k}\right]_{\text{equilibrium}} \tag{5-3}$$

This definition is the same as that of Eq. (4-1), but the description of a multicomponent system requires a number of such relative volatilities equal to the number of components in the system minus one. Combining Eqs. (5-1) and (5-3) yields

$$\alpha_{j,k} \equiv \frac{K_j}{K_k} \tag{5-4}$$

The order of the subscripts on α determines which component is placed in the numerator and which in the denominator in the right-hand sides of Eqs. (5-3) and (5-4).

In defining the relative volatility in a multicomponent system, it is conventional to pick a reference component and to define all relative volatilities with respect to that component. In this case, the second subscript is omitted from the symbol α because it is clearly understood that the second component is always the specified reference component. Then the relative volatility of any component with respect to this reference component is given by

$$\alpha_j \equiv \frac{K_j}{K_{\text{ref}}} \tag{5-5}$$

This equation defines the relative volatility of component j with respect to the reference component. The number of such relative volatility values which must be specified is equal to the number of components minus one. For the reference component itself, its relative volatility is obviously equal to unity.

Figure 5-2 shows the effect of temperature on the relative volatility values for propane, isobutane, *n*-butane, isopentane, *n*-pentane, and *n*-hexane at a pressure of 4.4 atm. Isopentane has been chosen as the reference component, and therefore all relative volatilities are expressed relative to

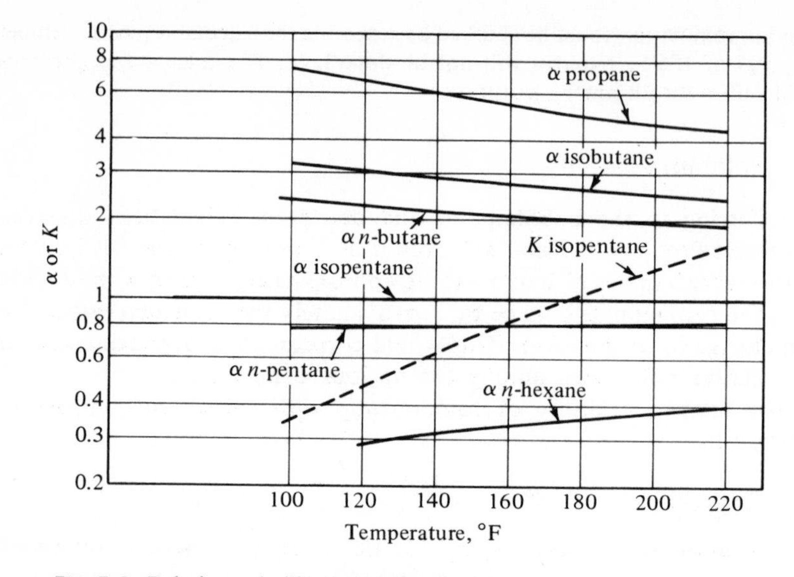

Fig. 5-2. Relative volatility values for light hydrocarbons at a pressure of 4.4 atm.

isopentane. Figure 5-2 was constructed from the values in Fig. 5-1 by simply dividing the K value for any component at a given temperature by the K value for isopentane at that same temperature. As is customary, the K value for the reference component, isopentane, is also shown in Fig. 5-2 for reference. Furthermore, for emphasis, the relative volatility of isopentane itself is shown in Fig. 5-2 as a horizontal line corresponding to a relative volatility of unity.

It is seen in Fig. 5-2 that the relative volatility values are not nearly so sensitive to temperature as are the K values shown in Fig. 5-1. This is especially true for the components adjacent in volatility to the reference component, *n*-butane and *n*-pentane for the case of Fig. 5-2. For a temperature change from 100°F to 200°F, the relative volatility values for *n*-pentane and *n*-butane change only 5 and 18 per cent, respectively. It is generally true in systems of this type that the percentage variation in the relative volatility of a component with temperature will increase as the volatility of that component departs more from the volatility of the reference component. Thus the choice of which component to employ as a reference component is sometimes influenced by the desire for the relative volatility values for certain key components to be the least sensitive to temperature.

The calculation of the bubble-point temperature and of the equilibrium-vapor composition for a specified liquid composition is straightforward when the relative volatility values are known or can be guessed with reasonable accuracy. Substitution of Eq. (5-5) into Eq. (5-1) yields the relationship between y and x for any component at equilibrium:

$$y_j = K_{\text{ref}}\alpha_j x_j \tag{5-6}$$

Equation (5-6) is written for each component, including the reference component, and these equations are summed to yield

$$\sum y_k = K_{\text{ref}} \sum \alpha_k x_k \tag{5-7}$$

where $\sum$ represents a summation on k for *all components*, including the reference component. But the sum of the vapor-phase mole fractions must be unity:

$$\sum y_k = 1 \tag{5-8}$$

Therefore, Eq. (5-7) becomes

$$K_{\text{ref}} = \frac{1}{\sum \alpha_k x_k} \tag{5-9}$$

Finally, substitution of Eq. (5-9) into Eq. (5-6) yields

$$y_j = \frac{\alpha_j x_j}{\sum \alpha_k x_k} \tag{5-10}$$

The bubble-point calculation proceeds by first computing the product of α and x for each component and summing the products for all the components. The reciprocal of this sum is, according to Eq. (5-9), the K value for the reference component. Division of the product of α and x for any component by the sum of such products for all components yields, according to Eq. (5-10), the equilibrium-vapor-phase mole fraction for that component. The bubble-point temperature is read as that temperature at which the reference component has the K value computed according to Eq. (5-9). If the relative volatility values do not vary with temperature, a single calculation will yield the equilibrium-vapor composition as well as the bubble-point temperature. When the relative volatility values vary with temperature, an iteration procedure is used in which the calculation is repeated if the relative volatility values assumed in the first place do not correspond closely to the actual relative volatility values at the temperature computed to be the bubble-point temperature.

For a specified vapor composition, the dew-point temperature and the equilibrium-liquid-composition calculations are also straightforward when the relative volatility values are known or can be guessed. Eq. (5-6) is rearranged to the form

$$x_j = \frac{y_j/\alpha_j}{K_{\text{ref}}} \tag{5-11}$$

This equation is written for each component, and the equations are then summed up to yield

$$\sum x_k = \frac{\sum y_k/\alpha_k}{K_{\text{ref}}} \tag{5-12}$$

But the liquid-phase mole fractions must add to unity:

$$\sum x_k = 1 \tag{5-13}$$

and so Eq. (5-12) becomes

$$K_{\text{ref}} = \sum \frac{y_k}{\alpha_k} \tag{5-14}$$

Finally, inserting Eq. (5-14) into Eq. (5-11) yields

$$x_j = \frac{y_j/\alpha_j}{\sum y_k/\alpha_k} \tag{5-15}$$

Thus the dew-point calculation proceeds by first dividing the vapor-phase mole fraction for each component by the relative volatility for that component and then by summing the quotients obtained for all components, including the reference component. This sum is equal to the K value for the reference component, according to Eq. (5-14). When the value of y/α for any component is divided by the sum of such values for all components, the result is equal to the liquid-phase mole fraction for that component, according to Eq. (5-15). The dew-point temperature is read as that temperature at which the reference component has the K value calculated by Eq. (5-14). Therefore, a straightforward calculation yields the dew-point temperature and the equilibrium-liquid composition when the relative volatility values can be guessed with sufficient accuracy. If the relative volatility values at the calculated dew-point temperature depart significantly from the values assumed at the beginning of the calculation, an iteration procedure is required.

These calculations will be illustrated with the following example problems.

EXAMPLE PROBLEM 5-3 Repeat Example Problem 5-1 using the relative volatility method.

Solution: As a first guess at the bubble-point temperature, 140°F will be chosen. At this temperature the relative volatility for each component can be read from Fig. 5-2. The value of x for each component is then multiplied by the value of α for that component, and the αx products are then summed up. Finally, this sum is divided into the αx product for each component to yield the vapor-phase mole fraction for that component. The calculations are summarized in Table 5-5.

Using Eq. (5-9) yields

$$K_{\text{ref}} = \frac{1}{1.605} = 0.623$$

The bubble-point temperature is read as that temperature at which the reference component, isopentane, has a K value of 0.623. From Fig. 5-2 or from Fig. 5-1, this temperature can be seen to be essentially 140°F. Since this temperature is in good agreement with the temperature assumed in the beginning, the calculation is

Table 5-5

Component	x	α *at 140°F*	αx	$y = \alpha x / \sum \alpha_k x_k$
Isobutane	0.2	2.87	0.574	0.3575
n-Butane	0.3	2.15	0.645	0.402
Isopentane	0.15	1.0	0.15	0.0935
n-Pentane	0.35	0.787	0.236	0.147
			1.605	1.0000

complete. The disagreement between this result and that obtained in Example Problem 5-1 is due to inaccuracies in reading Figs. 5-1 and 5-2.

EXAMPLE PROBLEM 5-4 Repeat Example Problem 5-2 using the relative volatility method.

Solution: As a first guess, 140°F will be chosen for the dew-point temperature. The relative volatility values at this temperature are read from Fig. 5-2, and then the value of y for each component is divided by its relative volatility. The y/α values are then summed up, and the y/α value for each component is then divided by the sum to yield the x value for that component. The calculations are summarized in Table 5-6.

Table 5-6

Component	y	α *at 140°F*	y/α	$x = \dfrac{y/\alpha}{\sum y_k/\alpha_k}$
Isobutane	0.25	2.87	0.0871	0.113
n-Butane	0.25	2.15	0.1163	0.151
Isopentane	0.25	1.0	0.25	0.324
n-Pentane	0.25	0.787	0.3175	0.412
			0.7709	1.000

Using Eq. (5-14), the K value for isopentane is

$$K_{\text{ref}} = 0.7709$$

The temperature at which the K value for isopentane is equal to 0.7709 is read from Fig. 5-2 as approximately 155°F. The calculation will now be repeated with relative volatility values read from Fig. 5-2 at a temperature of 155°F. The calculations are summarized in Table 5-7. This time, according to Eq. (5-14), the K value for the reference component is 0.7759. The temperature at which the K value for isopentane is equal to 0.7759 can be read from Fig. 5-2 as 155°F, within the accuracy with which the graph can be read. There is therefore no need to iterate the solution again. The dew-point temperature is 155°F, and the equilibrium-liquid composition is shown in the last column in Table 5-7. This result is reasonably close to that obtained in Example Problem 5-2, the difference being due to inaccuracies in reading the graphs.

Table 5-7

Component	y	α *at 155°F*	y/α	$x = \frac{y/\alpha}{\sum y_k/\alpha_k}$
Isobutane	0.25	2.75	0.0909	0.1172
n-Butane	0.25	2.075	0.1205	0.1553
Isopentane	0.25	1.0	0.25	0.3223
n-Pentane	0.25	0.795	0.3145	0.4052
			0.7759	1.0000

MULTICOMPONENT DISTILLATION-COLUMN SPECIFICATIONS

In the design of a binary distillation column, the compositions of the product streams to be obtained from a given feed stream can be specified uniquely. The column design then consists of determining the number of theoretical plates required to achieve the specified separation as a function of the reflux ratio employed. Exactly the same separation can be accomplished with a large number of plates and a small reflux ratio or with a smaller number of plates and a larger reflux ratio. In multicomponent distillation-column design, on the other hand, the compositions of the product streams cannot, in general, be specified completely. For a fractional-distillation column with a single feed stream, a distillate product stream, and a bottom product stream, it is generally possible to set only two specifications on the product stream compositions. Of course, two specifications can completely determine the compositions of two binary product streams, but two multicomponent product streams will generally not be uniquely determined by two specifications.

It is customary, therefore, in the specification of multicomponent separations to select two *key components* and to specify the distributions of these components between the distillate and bottom product streams. The distributions of the other components between the distillate and bottom product streams cannot then be specified but rather will be determined by the design conditions selected to accomplish the specified separation of the key components.

The two specifications to be made on the distribution of the key components between the distillate and bottom product streams may take several forms. Referring to the more volatile key component as the *light key* and to the less volatile key component as the *heavy key*, it is usual to specify the allowable mole fraction of the light key in the bottom product stream and the allowable mole fraction of the heavy key in the distillate product stream. Alternatively, one might wish to specify the allowable mole fraction of the heavy key in the distillate product stream and the recovery of the light-key component in the distillate product stream. Still another choice would be to specify the recovery of the light-key component in the distillate product and the recovery of the heavy key in the bottom product. A number of choices of this type are available, but the choice is not completely arbitrary. For example, if one were to specify the mole fractions of both key components in the distillate product stream, the specification might be acceptable, but it also might be that the values specified could not be achieved with any number of theoretical plates or any reflux ratio.

When the key components are adjacent in volatility and when the specified separation between the key components is relatively sharp, it is generally possible to determine the overall material balance around the entire distillation column in an approximate manner. The exact compositions in the distillate product of the components less volatile than the heavy key, called the *heavy components*, cannot be known exactly, but they can generally be assumed to be small. Similarly, exact mole fractions in the bottom product of the components more volatile than the light key, called the *light components*, cannot be known exactly, but they can be presumed to be quite small. These ideas are best illustrated by means of an example problem.

EXAMPLE PROBLEM 5-5 A multicomponent mixture contains propane, isobutane, *n*-butane, isopentane, *n*-pentane, and *n*-hexane with mole fractions of 0.05, 0.1, 0.3, 0.2, 0.15, and 0.2, respectively. It is desired to separate this mixture by fractional distillation into a distillate product stream with an isopentane mole fraction of 0.005 and a bottom product stream with an *n*-butane mole fraction of 0.003. With these specifications, determine the material balance as completely as possible.

Solution: The light-key component in this case is *n*-butane because its presence in the bottom product stream must be limited to a mole fraction of 0.003. The heavy-key component is isopentane because its presence in the distillate product stream must be limited to a mole fraction of 0.005. To make a material balance it will be assumed that the mole fractions in the distillate product of components less volatile than isopentane are very small compared to unity. It will likewise be assumed that the mole fractions in the bottom product of components more volatile than *n*-butane will be very small compared to unity. With these assumptions a material-balance table can be constructed on the basis of 100 moles of feed, as shown in Table 5-8.

The components are arranged in order of their volatilities, and the component names are abbreviated according to the usual convention of indicating the

Table 5-8

	Molal Flow Rate (moles/unit time)		
Component	*Feed*	*Distillate Product*	*Bottom Product*
C_3	5	5	~ 0
i-C_4	10	10	~ 0
n-C_4 (light key)	30	$30 - \epsilon$	ϵ
i-C_5 (heavy key)	20	μ	$20 - \mu$
n-C_5	15	~ 0	15
n-C_6	20	~ 0	20
Total	100	$45 - \epsilon + \mu$	$55 + \epsilon - \mu$

number of carbon atoms in the hydrocarbon molecule. Normal butane is the light-key component and isopentane is the heavy-key component. Propane and isobutane are light components, and the quantity of these components appearing in the bottom product is assumed to be negligible. The heavy components are n-pentane and n-hexane, and the quantity of these components appearing in the distillate product is assumed to be negligible. The small amount of n-butane appearing in the bottom product is represented by the unknown, ϵ, and the small amount of isopentane appearing in the distillate product is represented by μ. Since the specifications in this case are upon the mole fractions of the key components rather than upon recoveries, the required mole fractions are expressed in terms of the unknown quantities ϵ and μ:

$$\frac{\mu}{45 - \epsilon + \mu} = 0.005$$

$$\frac{\epsilon}{55 + \epsilon - \mu} = 0.003$$

Solving these two equations simultaneously yields

$$\mu = 0.2253, \qquad \epsilon = 0.1648$$

With these values, the material-balance table can be completed, as shown in Table 5-9. This table, of course, is based upon the assumption that the heavy components are present in the distillate product to a negligible extent and that the light components are present in the bottom product to a negligible extent. The quantity of n-pentane in the distillate product, for example, is actually not zero, but if it is a number that is very small compared to 15, then the fact that it is not zero will not affect the other numbers shown in the table to any appreciable extent. On the other hand, it is possible that the magnitude of the small amount of n-pentane in the distillate product might be of interest in the separation, even though isopentane was specified as the heavy-key component, but for now this magnitude is not known.

The molal flow rates in the table above could obviously be converted to mole fractions by simply dividing by the total molal flow rate of the stream in question. Therefore, this partial material balance provides reasonably accurate mole fractions for the key components and the light components in the distillate product and

Table 5-9

	Molal Flow Rate (moles/unit time)		
Component	*Feed*	*Distillate Product*	*Bottom Product*
C_3	5	5	0
i-C_4	10	10	0
n-C_4	30	29.8352	0.1648
i-C_5	20	0.2253	19.7747
n-C_5	15	0	15
n-C_6	20	0	20
		45.0605	54.9395

for the key components and the heavy components in the bottom product. Furthermore, reasonably acccurate values of the required distillate flow rate and of the required bottom product flow rate are obtained by this incomplete material balance.

5-4. Number of plates versus reflux ratio

As in binary distillation-column design, it is also true in multicomponent distillation-column design that an increase in the reflux ratio will result in a reduction in the number of theoretical plates required to achieve a specified separation. On the other hand, the specified separation is specified only in terms of the two key components, and as the reflux ratio is increased and the number of theoretical plates required is decreased, the unspecified concentrations in the product streams will generally not remain constant. For example, the mole fractions of the heavy components in the distillate product and of the light components in the bottom product will generally vary with the reflux ratio. As the reflux ratio approaches its minimum value and the number of theoretical plates approaches infinity, the mole fractions of the heavy components in the distillate product and of the light components in the bottom product will generally approach zero. As the reflux ratio is increased and the number of theoretical plates is decreased, the mole fractions of the heavy components in the distillate product and of the light components in the bottom product will generally increase, possibly passing through maxima before total reflux is approached. If it is indeed only the specifications on the key components that are of interest, then essentially the same separation can be achieved with a low reflux ratio and a large number of plates, or with a high reflux ratio and a smaller number of plates. But if the mole fractions of the nonkey components in the product streams are of any importance, then two such points on the design curve of theoretical plates required versus reflux ratio do not correspond to exactly the same separation.

APPROXIMATE DESIGN OF MULTICOMPONENT COLUMNS

The solution of the equations describing a multicomponent distillation column involves substantially more computational effort than does the solution for a binary distillation column. Furthermore, as will be seen later, the more effective methods for solving multicomponent distillation problems are *simulation* methods rather than design methods. That is, these methods determine what the product compositions will be when the number of plates, the feed-plate location, the reflux ratio, and the distillate and bottom product flow rates are all specified. To use these methods to design a distillation column to achieve a specified separation, a trial-and-error procedure is used. For these reasons it is especially useful to have simple approximate methods of designing the multicomponent system to enable preliminary economic balance studies to be made and to guide the trial-and-error selection of the proper reflux ratio, which is then checked by the simulation program.

A number of approximate multicomponent distillation-column-design procedures have been proposed in the literature. In this introductory treatment, only the use of the Gilliland correlation [4, 13] will be considered. This procedure is perhaps the simplest, and its accuracy is usually adequate for preliminary economic considerations and for guiding the use of an exact simulation program.

The Gilliland correlation was presented in Chap. 4 as Fig. 4-48. This correlation is applicable to multicomponent distillation columns as well as to binary distillation columns. As presented, the correlation predicts the entire curve of the number of theoretical plates required versus the reflux ratio employed in terms of only two parameters, the minimum reflux ratio and the minimum number of theoretical plates, corresponding to total reflux. As indicated in Chap. 4, a more accurate correlation could be obtained by using different curves for different feed compositions, different values of q, and perhaps also for different values of the relative volatility. But the use of a single curve provides a simple correlation whose accuracy is sufficient for most cases. As in the case of a binary distillation column, the minimum reflux ratio and the minimum number of plates must be determined before the Gilliland correlation can be employed.

5-5. Total reflux

Equations (4-10) and (4-12) represent the material-balance relationships for the stripping section and the enriching section, respectively, for a multicomponent distillation column as well as for a binary distillation column. For the multicomponent case, equations of this type are written for each component. As in the case of a binary distillation column, as the reflux ratio approaches infinity, the material-balance relationships in both the stripping and the enriching sections of the column approach the form

$$x_{j,n+1} = y_{j,n} \tag{5-16}$$

The first subscript on the mole fractions represents the component, component j in this case, and the second subscript represents the number of the plate from which this stream is leaving. Thus, for component j, Eq. (5-16) states that the liquid overflow from plate $n + 1$ has the same composition as the vapor rising from plate n, in the limit as the reflux ratio approaches infinity. This equation applies for any component. For the case in which the relative volatilities among the various components are constant or can be assumed to be constant as an approximation, the equilibrium relationship of Eq. (5-3) can be applied to each plate in the column to give

$$\frac{y_{j,n}}{x_{j,n}} = \alpha_{j,k}\frac{y_{k,n}}{x_{k,n}} \tag{5-17}$$

This equation can be written for any pair of components, j and k.

Using Eqs. (5-16) and (5-17), the Fenske equation analogous to Eq. (4-43) can be developed for a multicomponent distillation column at total reflux. Attention is focused upon any pair of components, j and k. Assuming that the vapor from the reboiler is in equilibrium with the bottom product composition, Eq. (5-17) yields

$$\frac{y_{j,0}}{x_{j,B}} = \alpha_{j,k}\frac{y_{k,0}}{x_{k,B}} \tag{5-18}$$

Equation (5-16) states that the composition of the liquid leaving the first plate is equal to the composition of the vapor from the reboiler. Employing this relationship for both components j and k, Eq. (5-18) becomes

$$\frac{x_{j,1}}{x_{j,B}} = \alpha_{j,k}\frac{x_{k,1}}{x_{k,B}} \tag{5-19}$$

The equilibrium relationship for the first plate is

$$\frac{y_{j,1}}{x_{j,1}} = \alpha_{j,k}\frac{y_{k,1}}{x_{k,1}} \tag{5-20}$$

Equation (5-16) requires that the composition of the liquid overflow from the second plate must be equal to the composition of the vapor leaving the first plate. Employing this relationship for both components j and k, Eq. (5-20) becomes

$$\frac{x_{j,2}}{x_{j,1}} = \alpha_{j,k}\frac{x_{k,2}}{x_{k,1}} \tag{5-21}$$

Multiplying Eq. (5-19) by Eq. (5-21) yields

$$\frac{x_{j,2}}{x_{j,B}} = (\alpha_{j,k})^2 \frac{x_{k,2}}{x_{k,B}} \tag{5-22}$$

The equilibrium relationship for the second plate is

$$\frac{y_{j,2}}{x_{j,2}} = \alpha_{j,k} \frac{y_{k,2}}{x_{k,2}} \tag{5-23}$$

Combining this equation with Eq. (5-16) for $n = 2$ yields

$$\frac{x_{j,3}}{x_{j,2}} = \alpha_{j,k} \frac{x_{k,3}}{x_{k,2}} \tag{5-24}$$

Multiplying Eq. (5-24) by Eq. (5-22) yields

$$\frac{x_{j,3}}{x_{j,B}} = (\alpha_{j,k})^3 \frac{x_{k,3}}{x_{k,B}} \tag{5-25}$$

Continuing this process up to the top of the column gives

$$\frac{x_{j,N}}{x_{j,B}} = (\alpha_{j,k})^N \frac{x_{k,N}}{x_{k,B}} \tag{5-26}$$

Multiplying this equation by Eq. (5-17) written for $n = N$ gives

$$\frac{y_{j,N}}{x_{j,B}} = (\alpha_{j,k})^{N+1} \frac{y_{k,N}}{x_{k,B}} \tag{5-27}$$

Assuming that a total condenser is employed, the composition of the vapor leaving the top plate is equal to the distillate product composition, and Eq. (5-27) becomes

$$\frac{x_{j,D}}{x_{j,B}} = (\alpha_{j,k})^{N+1} \frac{x_{k,D}}{x_{k,B}} \tag{5-28}$$

This equation can be written in a form analogous to Eq. (4-42):

$$\mathcal{S}_{j,k} \equiv \frac{x_{j,D}/x_{k,D}}{x_{j,B}/x_{k,B}} \equiv \frac{Dx_{j,D}/Bx_{j,B}}{Dx_{k,D}/Bx_{k,B}} = (\alpha_{j,k})^{N+1} \tag{5-29}$$

Equation (5-29), first derived by Fenske [2], can be applied to any pair of components, j and k, in a multicomponent mixture. For the special case of a binary mixture, this equation reduces to Eq. (4-42). In applying the Fenske equation to a multicomponent distillation column, Eq. (5-29) is generally written first for the key components and solved for the number of stages required at total reflux. Then, after the number of stages has been determined, the equation can be applied to a pair of components consisting of one key component and one nonkey component to determine the concentration of the nonkey component in the product streams at total reflux. The equation is exact when applied to any pair of components whose relative volatility is constant; it is useful as an approximation in any situation in which the variation in the relative volatility of the two components is not too great. As for binary distillation, the exponent $N + 1$ is replaced by $N + 2$ when an equilibrium partial condenser is employed.

5-6. Minimum reflux ratio

The determination of the minimum reflux ratio for a multicomponent distillation separation is much more difficult than it is for a binary separation. Underwood's method [15] is exact for the case of constant molal overflow and constant relative volatilities. Gilliland's method [3, 11] establishes limits on the minimum reflux ratio which are usually sufficiently close to each other to define the minimum reflux ratio with good accuracy, even when the relative volatilities vary substantially. Colburn's method [1] is similar but uses an empirical correlation to establish the position of the answer within the range established by the limiting values. But when the relative volatilities are not constant, these last two methods involve considerable calculational difficulty. Gilliland's method has the great advantage, however, that it imparts considerable insight into the behavior of a multicomponent distillation column near the condition of minimum reflux. These methods are highly recommended for future study, but in this textbook only the Underwood method will be considered.

Underwood obtained an exact solution to the equations describing a multicomponent distillation column with constant molal overflow and with constant relative volatilities among all components. For the rather complex mathematical derivation, the reader is referred to Underwood's original paper [15]. Here the equations will simply be given and their use illustrated. These equations are limited to the case of a single feed stream and two product streams, a distillate product and a bottom product stream.

The Underwood equations for minimum reflux are

$$\sum \frac{\alpha_k z_{k,F}}{\alpha_k - \phi} = 1 - q \tag{5-30}$$

$$\mathscr{R}_{\text{min}} + 1 = \sum \frac{\alpha_k x_{k,D}}{\alpha_k - \phi} \tag{5-31}$$

The symbol $\sum$ indicates a summation on the index k for all components. The values of α_k, $z_{k,F}$, and $x_{k,D}$ are, in general, different for different components, but the parameter ϕ refers to a single value.

When the multicomponent separation is a reasonably sharp split between key components adjacent in volatility, accurate values of x_D can be calculated for all components except the heavy components. In this case, the Underwood equations are employed by first solving Eq. (5-30) for the value of ϕ lying between the relative volatilities of the key components and then substituting this value of ϕ into Eq. (5-31) to yield the value of $\mathscr{R}_{\text{min}}$. It must be emphasized that there are a number of roots to Eq. (5-30), corresponding to ϕ values between each pair of adjacent α values, but the desired root is that one lying between the α values for the key components. When this value of ϕ is substituted into Eq. (5-31), the x_D values for the heavy components will generally be taken to be zero.

Occasionally, the key components are selected such that there is a component of volatility intermediate between the key components, called a *distributed component*. In this case, there are two roots of Eq. (5-30) lying between the α values for the key components. Underwood [16] presents a method for handling such cases, but this will not be considered here.

The approximate design of a multicomponent distillation column can best be illustrated by means of an example problem.

EXAMPLE PROBLEM 5-6 A saturated liquid mixture with the same composition as the feed mixture in Example Problem 5-5 is to be separated into a distillate product containing 0.5 mole per cent isopentane and a bottom product containing 0.5 mole per cent n-butane. The column will operate at a pressure of 4.4 atm and will be equipped with a partial reboiler and a total condenser. Estimate the design curve of the number of theoretical plates required versus the reflux ratio employed.

Solution: The partial material balance can be made just as it was in Example Problem 5-5, except that here the contamination of the bottom product with the light-key component, n-butane, is 0.5 mole per cent rather than 0.3 mole per cent as in Example Problem 5-5. The partial material balance thus determined is summarized in Table 5-10. These molal flow rates are readily converted to mole frac-

Table 5-10

	Molal Flow Rate (moles/unit time)		
Component	*Feed*	*Distillate Product*	*Bottom Product*
C_3	5	5	~0
i-C_4	10	10	~0
n-C_4(light key)	30	29.7248	0.2752
i-C_5 (heavy key)	20	0.2247	19.7753
n-C_5	15	~0	15
n-C_6	20	~0	20
Total	100	44.9495	55.0505

tions in the distillate and bottom product streams, as in Table 5-11. The composition of the vapor leaving the top plate in the column is equal, of course, to the distillate product composition. The dew-point temperature of this vapor is the temperature on the top plate, and the bubble-point temperature of the bottom liquid product is the temperature in the reboiler. Using the method illustrated in Example Problems 5-3 and 5-4, these temperatures are calculated to be 103°F for the top plate temperature and 206°F for the reboiler temperature.

Table 5-11

Component	Mole Fraction in	
	Distillate Product	*Bottom Product*
C_3	0.1113	~0
i-C_4	0.2226	~0
n-C_4	0.6611	0.005
i-C_5	0.005	0.359
n-C_5	~0	0.2725
n-C_6	~0	0.3635
	1.0000	1.0000

The relative volatility of n-butane with respect to isopentane is read from Fig. 5-2 as 2.34 at 103°F and 1.93 at 206°F. It would appear reasonable to assume a constant relative volatility of 2.1 as an approximation. Applying Eq. (5-29) to the light- and heavy-key components, n-butane and isopentane, results in

$$\frac{29.7248/0.2752}{0.2247/19.7753} = (2.1)^{N+1}$$

Solving this equation for the minimum number of stages at total reflux gives

$$\mathcal{N}_{\text{min}} = N_{\text{min}} + 1 = 12.35$$

Equation (5-29) can now be used to estimate the distribution of the nonkey components between the distillate and bottom product streams at total reflux. Consider isobutane first. The relative volatility of isobutane with respect to isopentane is 3.18 at 103°F and 2.47 at 206°F. Assuming a constant value of this relative volatility equal to 2.8, Eq. (5-29) can be written for isobutane and isopentane to give

$$\frac{10/Bx_{i-C_4,B}}{0.2247/19.7753} = (2.8)^{12.35}$$

Solving this equation yields

$$Bx_{i-C_4,B} = 0.0026$$

and dividing by the bottom product flow rate gives

$$x_{i-C_4,B} = \frac{0.0026}{55.0505} = 4.72 \times 10^{-5}$$

It can now be seen that the molal flow rate of isobutane in the bottom product stream is indeed quite small compared to the isobutane feed rate and also compared to the distillate and bottom product flow rates.

In a similar manner Eq. (5-29) can be used to determine the quantity of propane flowing in the bottom product stream. Employing an average relative volatility of propane with respect to isopentane of 5.6, the result is

$$Bx_{C_3,B} = 2.6 \times 10^{-7}, \qquad x_{C_3,B} = 4.7 \times 10^{-9}$$

The relative volatility of n-pentane with respect to isopentane varies from

0.775 at 103°F to 0.815 at 206°F, and therefore an average value of 0.8 will be taken. Applying Eq. (5-29) to n-pentane and isopentane yields

$$\frac{Dx_{n-C_5,D}/15}{0.2247/19.7753} = (0.8)^{12.35}$$

Solving this equation gives

$$Dx_{n-C_5,D} = 0.0108$$

and dividing this by the distillate product flow rate gives

$$x_{n-C_5,D} = 0.00024$$

It can now be seen that the flow rate of n-pentane in the distillate product is approximately $\frac{1}{20}$ the flow rate of isopentane in the distillate product, but this quantity is still quite small relative to the n-pentane feed rate and to the distillate and bottom product flow rates; therefore, the approximate material balance made by neglecting this quantity is not in serious error.

Taking the average relative volatility of n-hexane with respect to isopentane to be 0.33, the flow rate of n-hexane in the distillate product is computed in a similar manner to be

$$Dx_{n-C_6,D} = 2.6 \times 10^{-7}, \quad x_{n-C_6,D} = 5.8 \times 10^{-9}$$

The preceding calculations illustrate the use of Eq. (5-29). First, this equation is employed for the two key components to determine the minimum number of stages corresponding to total reflux. Next, Eq. (5-29) is employed with one nonkey and one key component to determine the distribution of the nonkey component between the distillate and bottom product streams at total reflux. In this manner, refinements to the approximate material balance, which neglected the light components in the bottom product and the heavy components in the distillate product, can be made for an infinite reflux ratio. At finite reflux ratios the quantities of the light components in the bottom product and the heavy components in the distillate product will be somewhat different from the corresponding values for an infinite reflux ratio.

Next the minimum reflux ratio for this multicomponent separation will be calculated. Assuming the same constant relative volatilities as those employed with the Fenske equation, Eq. (5-30) becomes

$$\frac{(5.6)(0.05)}{5.6 - \phi} + \frac{(2.8)(0.1)}{2.8 - \phi} + \frac{(2.1)(0.3)}{2.1 - \phi} + \frac{(1.0)(0.2)}{1.0 - \phi} + \frac{(0.8)(0.15)}{0.8 - \phi} + \frac{(0.33)(0.2)}{0.33 - \phi} = 0$$

since $q = 1$ for this saturated liquid feed. It is now necessary to find the root of the above equation lying between 1.0 and 2.1. Denoting the left-hand side of the equation $f(\phi)$, first $f(\phi)$ is calculated at an initial guess $\phi = 1.5$. The result is $f(\phi) = 0.706$. Inspection of the above equation reveals that $f(\phi)$ increases with increasing ϕ, and so a smaller value of ϕ is tried next. At $\phi = 1.3$, $f(\phi)$ is computed to be 0.0644. Plotting these two values, as shown in Fig. 5-3, suggests that $\phi = 1.28$ is a good next guess. At $\phi = 1.28$, $f(\phi)$ is found to be -0.0163. At $\phi = 1.284$, $f(\phi) = 0$ within the accuracy of a slide-rule calculation, and thus the desired root is $\phi = 1.284$.

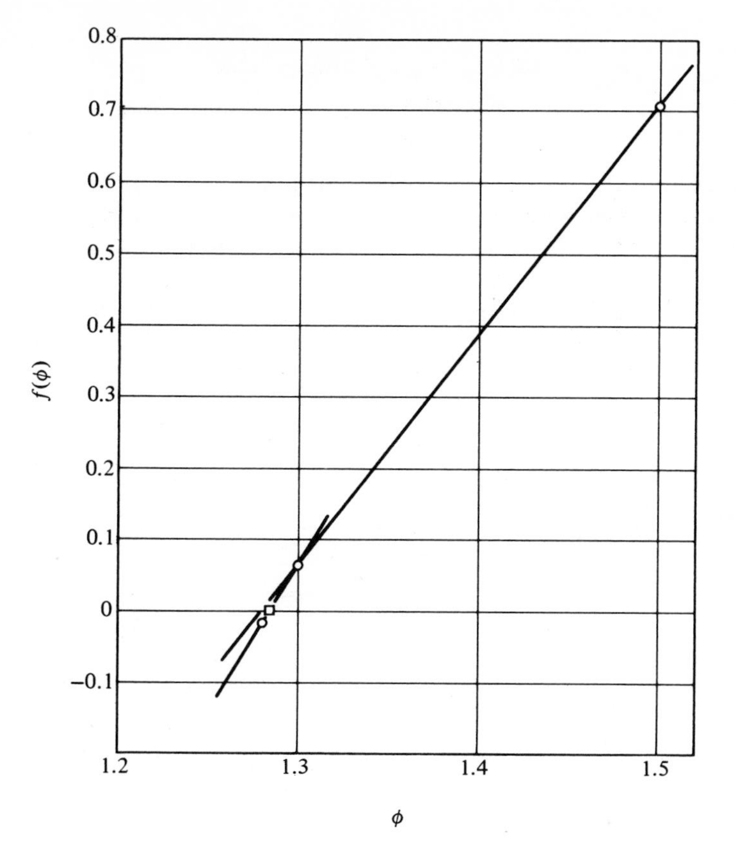

Fig. 5-3. Graphical solution for ϕ.

Substituting this root into Eq. (5-31) yields the minimum reflux ratio

$$\mathscr{R}_{\min} + 1 = \frac{(5.6)(0.1113)}{5.6 - 1.284} + \frac{(2.8)(0.2226)}{2.8 - 1.284} + \frac{(2.1)(0.6611)}{2.1 - 1.284} + \frac{(1.0)(0.005)}{1.0 - 1.284}$$
$$= 2.24$$

using $x_D = 0$ for the heavy components, *n*-pentane and *n*-hexane. Therefore,

$$\mathscr{R}_{\min} = 1.24$$

Using the values $\mathscr{R}_{\min} = 1.24$ and $\mathscr{N}_{\min} = 12.35$, the Gilliland correlation is next used to estimate the design curve. Consider, for example, a reflux ratio of 1.6. The abscissa of the Gilliland correlation is

$$\frac{\mathscr{R} - \mathscr{R}_{\min}}{\mathscr{R} + 1} = \frac{1.6 - 1.24}{1.6 + 1} = 0.1385$$

Reading Fig. 4-48 at this abscissa value yields

$$\frac{\mathscr{N} - \mathscr{N}_{\min}}{\mathscr{N} + 1} = 0.495$$

Thus

$$0.505\mathcal{N} = \mathcal{N}_{\min} + 0.495 = 12.85$$

$$\mathcal{N} = 25.5$$

$$N = 24.5$$

The Gilliland correlation therefore predicts that $N = 24.5$ when $\mathcal{R} = 1.6$. Similar calculations at other reflux ratios yield the values in the estimated design curve for this multicomponent distillation-column design, as shown in Fig. 5-4.

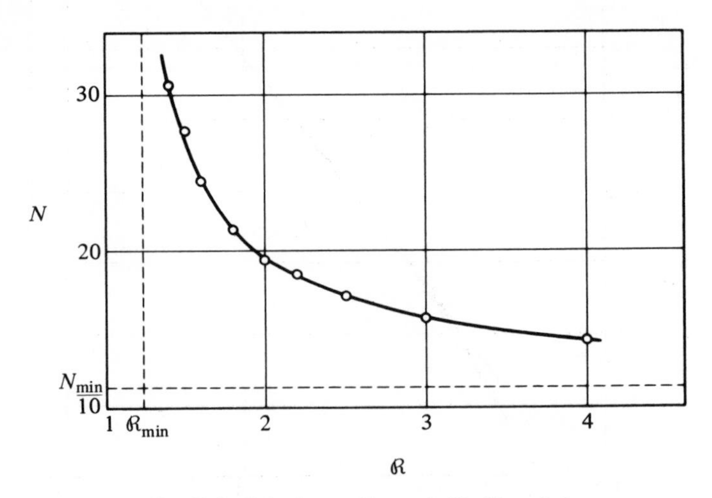

Fig. 5-4. Solution to Example Problem 5-6.

MULTICOMPONENT DISTILLATION-COLUMN SIMULATION AND DESIGN

The simultaneous solution of the many equations describing the material balances and the vapor–liquid equilibrium relationships in a multicomponent distillation column is much more difficult than it is for a binary distillation column. The Lewis–Matheson method [6, 9], a plate-to-plate design calculation analogous to that employed in binary distillation-column design, is quite satisfactory for a multicomponent separation containing either light components or heavy components, but this type of calculation is complicated enormously when the multicomponent separation involves both light and heavy components. In this general case, it is usually preferable to approach the problem by a different mathematical technique involving an iterative solution to the equations describing the simulation of a multicomponent distillation column.

5-7. Lewis–Matheson method

Consider a multicomponent separation involving components less volatile than the heavy-key component but involving no components more volatile than the light-key component. Because of the absence of light components, the material balance will determine the bottom product composition completely, provided that the separation between the key components is reasonably sharp. In this case, a plate-to-plate design calculation can be started at the reboiler and carried right up the column in a manner analogous to the binary plate-to-plate calculation described in Chap. 4 and illustrated in Example Problems 4-1 and 4-2.

Knowing the bottom product composition and assuming that an equilibrium partial reboiler is used, the composition of the vapor entering the first plate can be calculated as that composition in equilibrium with the bottom product liquid, employing the methods of Example Problem 5-1 or 5-3. The composition of the liquid leaving the bottom plate is then calculated by a material balance. Assuming constant molal overflow, Eqs. (4-7), (4-9), and (4-11) apply to a multicomponent distillation column as well as to a binary distillation column. Equations (4-8), (4-10), and (4-12) also apply to a multicomponent distillation column, but these equations must be written for each component in the multicomponent system. Thus, after determining the composition of the vapor leaving the reboiler, the composition of each component in the liquid leaving the bottom plate is readily computed by Eq. (4-10), with $m = 0$, written separately for each component. Equation (4-10) can be applied to each component because the mole fraction of each component in the bottom product is known with good accuracy.

After the complete composition of the liquid overflow from the first plate has been determined, the composition of the vapor rising from the first plate is then calculated as that vapor in equilibrium with the liquid flowing from the first plate. This calculation can be accomplished by the methods of Example Problem 5-1 or 5-3. Next, the complete composition of the liquid overflow from the second plate is obtained by application of Eq. (4-10), with $m = 1$, written for each component in the multicomponent system. This plate-to-plate calculation is continued up the column in a manner entirely analogous to that employed in a binary distillation calculation. After the feed is introduced onto plate f, Eq. (4-12), the enriching-section material balance, is used for each component for the calculation of the composition of the liquid overflow from plate $f + 1$ and for subsequent plates up the column.

Except for the different calculational method for obtaining the equilibrium-vapor composition for a given liquid composition, the plate-to-plate calculation for the multicomponent separation with no light components is entirely analogous to that for a binary separation. The determination of the optimum feed-plate location, which will minimize the total number of plates, is, however, somewhat complicated. As in the binary case, a trial-and-error procedure can be employed where, on any given plate, the composition of the liquid from the plate above can be calculated with both Eqs. (4-10) and (4-12). But, unlike the binary case, it will not always be apparent which equation gave the greatest enhancement in the separation and therefore whether or not it is advantageous to introduce the feed at that point. The safest way is to introduce the feed onto several different plates and for each choice of feed-plate location to carry the calculation all the way up to the top plate. It will then be a simple matter to see which choice of feed-plate location resulted in the minimum total number of plates required for the separation.

After the feed is introduced, the enriching-section material balance, Eq. (4-12), is employed for the plate-to-plate calculation up the column. In Eq. (4-12), x_D is taken to be zero for those components less volatile than the heavy-key component. This approximation will not impair the accuracy of the calculation provided that the actual values of x_D for these heavy components are indeed quite small. Even then, however, the choice of $x_D = 0$ in Eq. (4-12) will introduce some error into the calculation of the x values for the heavy components on the top plate and perhaps also on the plate beneath. If accurate values of x_D for the heavy components are desired, an iteration procedure can be used. First, the value of x_D for each of the heavy components is determined by carrying the calculation up to the top of the column while employing $x_D = 0$ for these components in Eq. (4-12). Next, the calculation for the top plate and perhaps one or two plates beneath is repeated, employing in Eq. (4-12) the values of x_D obtained in the first calculation. The iteration can then be repeated if necessary.

For a multicomponent separation in which there are light components but no heavy components, the composition of the distillate product can be known with good accuracy, assuming that the separation between the key components is a sharp one. In this case, a plate-to-plate calculation can be carried down the column beginning with the distillate composition. Assuming a total condenser, the vapor leaving the top plate has the same composition as the distillate product. The complete composition of the liquid leaving the top plate is calculated as that liquid in equilibrium with the vapor leaving the top plate. This calculation can be accomplished by the methods of Example Problem 5-2 or 5-4. Knowing the complete composition of the liquid overflow from the top plate, the composition of the vapor rising from the plate beneath the top plate can be calculated by using Eq. (4-12), written for each component in the multicomponent mixture. This equation can be applied because accurate values of x_D are known for all components. The plate-to-plate calculation is then carried down the column, first using a dew-point calculation to determine the liquid composition in equilibrium with the vapor rising from a given plate, and then employing Eq. (4-12) to calculate the composition of the vapor rising from the plate beneath. After the feed is introduced, the stripping-section material balance, Eq. (4-10), is employed. In Eq. (4-10) the x_B values for the light components are taken to be zero. This will introduce negligible error except in the calculation of the values of y for the light components in the vapor from the reboiler and perhaps for the first few plates. More accurate values of these vapor compositions can be obtained by repeating the calculation for the bottom few plates employing in Eq. (4-10) values of x_B for the light components which were obtained in the first calculation. The optimum feed-plate location is best determined by choosing several feed-plate locations and, with each choice, carrying the calculation all the way to the bottom of the column in order that the number of theoretical plates required with each choice of feed-plate location can be determined. Robinson and Gilliland [10] suggest guidelines for picking the optimum feed-plate location, but these will not be considered here.

It is seen from the preceding discussion that a plate-to-plate calculation can be carried up the column if no light components are present or down the column if no heavy components are present. But if a multicomponent separation involves both light and heavy components, the plate-to-plate calculation is seriously complicated. The problem arises from the fact that the concentration of a light component in the bottom product stream is unknown, and the concentration of a heavy component in the distillate product is unknown. Furthermore, it is not practical to attempt to guess the concentration of a light component in the bottom product stream and to then carry a plate-to-plate calculation up the column on the basis of such a guess. If this were done, it would be found that the mole fraction of the light component would build up very rapidly as the calculation proceeded up the column, and a very small error in the initial guess would result in a very large error in the mole fraction of that component farther up the column. A similar problem is encountered with the heavy component when calculating down the column, beginning with the distillate product composition. Techniques, called *feed-plate matching techniques*, have been developed to allow calculations up the column and calculations down the column to converge at the feed plate, but these will not be considered here. The reader is referred to other textbooks [5, 12] for a discussion of such techniques. Generally speaking, they are not as well suited to automatic computation as is the distillation-column-simulation technique discussed below.

5-8. Thiele–Geddes method

The plate-to-plate calculation described above is a design method because it is used to determine the number of theoretical plates required to achieve a specified separation at a specified reflux ratio. In contrast, the method of Thiele and Geddes [14] is a simulation method. It is used to determine the compositions of the distillate and bottom product streams when the total number of plates in the column, the feed-plate location, z_F, q, D/F, and the reflux ratio are all specified. Repeated application of a simulation method can, of course, yield a distillation-column design. For example, if a given column is simulated and it is found that the product purities do not meet the specifications desired, a higher reflux ratio can be selected for the same number of theoretical plates and the simulation can be repeated until the specifications are met. The Thiele–Geddes simulation method will be developed here for a multicomponent column with a single feed stream and with no side product streams.

Before beginning the Thiele–Geddes calculation the following variables must be specified: the number of theoretical plates in the column, the location of the feed plate, the composition and the q value of the feed stream, the ratio of the distillate or the bottom product flow rate to the feed rate, and the reflux ratio or the reboil ratio. The feed rate may be specified, or an arbitrary basis such as 100 moles of feed may be chosen. The flow rates of the distillate and bottom product streams and the liquid and vapor streams within the stripping and enriching sections of the column are then calculated. An iterative sequence of calculations is then used to determine the mole fractions of all components in all the liquid and vapor streams throughout the column.

The calculations are begun by assuming a temperature profile throughout the column. The temperature in the reboiler and the temperature on the top plate can usually be calculated with good accuracy, and one possible initial temperature guess is a linear variation in temperature with plate number between these limiting temperatures. But in any event some initial guess of a temperature distribution throughout the column is made. From this assumed temperature distribution, the K value for each component on each plate is determined. In terms of these K values, the equilibrium expression relating the vapor and liquid compositions leaving any plate is given simply by Eq. (5-1). The material-balance relationships are given by Eqs. (4-10) and (4-12) for the stripping and enriching sections of the column, respectively. These equations are written for each component in a multicomponent mixture.

It is convenient to define a symbol for the molal flow rate of component j in the liquid-overflow stream leaving plate m, any plate in the stripping section of the column:

$$\ell_{j,m} \equiv L'x_{j,m} \qquad (\text{for } m = 1, 2, \ldots, f) \tag{5-32}$$

and to define a symbol for the molal flow rate of component j in the vapor stream rising from any plate in the stripping section of the column, including the vapor stream from the reboiler:

$$\mathscr{V}_{j,m} \equiv V'y_{j,m} \qquad (\text{for } m = 0, 1, 2, \ldots, f-1) \tag{5-33}$$

It is also convenient to define a symbol for the molal flow rate of component j in the bottom product stream:

$$\mathscr{L}_j \equiv Bx_{j,B} \tag{5-34}$$

Assuming that a partial reboiler is used and that the vapor leaving the reboiler is in equilibrium with the bottom product liquid, the equilibrium relationship becomes

$$y_{j,0} = K_{j,R}x_{j,B} \tag{5-35}$$

Multiplying Eq. (5-35) through by V' and then multiplying and dividing the right-hand side by B yields

$$V'y_{j,0} = \frac{V'K_{j,R}}{B}Bx_{j,B} \tag{5-36}$$

Defining a *stripping factor* for component j in the reboiler as

$$S_{j,R} \equiv \frac{V'K_{j,R}}{B} \tag{5-37}$$

Eq. (5-36) can be rewritten in terms of the defined variables as

$$\mathscr{V}_{j,0} = S_{j,R}\mathscr{L}_j \tag{5-38}$$

The equilibrium relationship for component j on any plate in the stripping section of the column is

$$y_{j,m} = K_{j,m}x_{j,m} \qquad (\text{for } m = 1, 2, \ldots, f-1) \tag{5-39}$$

where $K_{j,m}$ is the K value for component j at the temperature of plate m. Multiplying Eq. (5-39) through by V' and then multiplying and dividing the right-hand side by L' yields

$$V'y_{j,m} = \frac{V'K_{j,m}}{L'}L'x_{j,m} \tag{5-40}$$

Defining a stripping factor for component j on plate m as

$$S_{j,m} \equiv \frac{V'K_{j,m}}{L'} \qquad (\text{for } m = 1, 2, \ldots, f-1) \tag{5-41}$$

Eq. (5-40) can be rewritten in terms of these defined quantities as

$$\mathscr{V}_{j,m} = S_{j,m}\ell_{j,m} \qquad (\text{for } m = 1, 2, \ldots, f-1) \tag{5-42}$$

The stripping-section material balance is given by Eq. (4-10), written for component j:

$$L'x_{j,m+1} = V'y_{j,m} + Bx_{j,B} \qquad (\text{for } m = 0, 1, \ldots, f-1) \tag{5-43}$$

In terms of the quantities defined, this equation can be rewritten as

$$\ell_{j,m+1} = \mathscr{V}_{j,m} + \mathscr{L}_j \qquad (\text{for } m = 0, 1, \ldots, f-1) \tag{5-44}$$

Since the Thiele–Geddes method is a simulation method, the bottom product composition is not known, and thus $\mathscr{L}_j$ is not known. Therefore, $\ell/\mathscr{L}$ and $\mathscr{V}/\mathscr{L}$ are employed as the variables in the solution of these equations.

Equations (5-38), (5-42), and (5-44) are divided through by $\mathscr{b}_j$ to give

$$\frac{\mathscr{V}_{j,0}}{\mathscr{b}_j} = S_{j,R} \tag{5-45}$$

$$\frac{\mathscr{V}_{j,m}}{\mathscr{b}_j} = S_{j,m}\frac{\ell_{j,m}}{\mathscr{b}_j} \qquad (\text{for } m = 1, 2, \ldots, f-1) \tag{5-46}$$

$$\frac{\ell_{j,m+1}}{\mathscr{b}_j} = 1 + \frac{\mathscr{V}_{j,m}}{\mathscr{b}_j} \qquad (\text{for } m = 0, 1, 2, \ldots, f-1) \tag{5-47}$$

When the K value is known for each component on each plate, since B, V', and L' are all known, the stripping factor can be calculated for each component on each plate in the stripping section of the column. For any component, a calculational procedure is then started at the reboiler and carried up the column to the feed plate as follows. First, $\mathscr{V}_{j,0}/\mathscr{b}_j$ is calculated from Eq. (5-45), using the known stripping factor in the reboiler. This value is then substituted into Eq. (5-47) with $m = 0$ to give $\ell_{j,1}/\mathscr{b}_j$. This value is then substituted into Eq. (5-46) with $m = 1$ to give $\mathscr{V}_{j,1}/\mathscr{b}_j$. This value is then substituted into Eq. (5-47) with $m = 1$ to give $\ell_{j,2}/\mathscr{b}_j$. This value is then substituted into Eq. (5-46) with $m = 2$ to give $\mathscr{V}_{j,2}/\mathscr{b}_j$. The alternate use of Eqs. (5-47) and (5-46) is then carried up the column to calculate all the $\ell/\mathscr{b}$ and $\mathscr{V}/\mathscr{b}$ values up to and including $\mathscr{V}_{j,f-1}/\mathscr{b}_j$ and $\ell_{j,f}/\mathscr{b}_j$.

A similar calculation is carried from the top of the column down to the feed plate. The molal flow rates of component j in the liquid and vapor streams leaving plate n, any plate in the enriching section of the column, are defined as

$$\ell_{j,n} \equiv Lx_{j,n} \qquad (\text{for } n = f+1, f+2, \ldots, N) \tag{5-48}$$

$$\mathscr{V}_{j,n} \equiv Vy_{j,n} \qquad (\text{for } n = f, f+1, \ldots, N) \tag{5-49}$$

and the molal flow rate of component j in the distillate product stream is defined as

$$\mathscr{d}_j \equiv Dx_{j,D} \tag{5-50}$$

The material-balance relationship for the enriching section of the column is given by Eq. (4-12), written for component j:

$$Vy_{j,n} = Lx_{j,n+1} + D_{j,D} \qquad (\text{for } n = f, f+1, \ldots, N) \tag{5-51}$$

This equation can be written in the form

$$\frac{\mathscr{V}_{j,n}}{\mathscr{d}_j} = 1 + \frac{\ell_{j,n+1}}{\mathscr{d}_j} \qquad (\text{for } n = f, f+1, \ldots, N) \tag{5-52}$$

The equilibrium relationship for any plate in the enriching section is

$$y_{j,n} = K_{j,n}x_{j,n} \qquad (\text{for } n = f+1, f+2, \ldots, N) \tag{5-53}$$

Defining an *absorption factor* for component j on plate n as

$$A_{j,n} \equiv \frac{L}{VK_{j,n}} \qquad (\text{for } n = f+1, f+2, \ldots, N) \tag{5-54}$$

Eq. (5-53) can be rearranged to the form

$$\frac{\ell_{j,n}}{\mathscr{d}_j} = A_{j,n}\frac{\mathcal{V}_{j,n}}{\mathscr{d}_j} \qquad (\text{for } n = f+1, f+2, \ldots, N) \tag{5-55}$$

The calculation is carried down the column as follows. Assuming that a total condenser is used,

$$y_{j,N} = x_{j,D} \tag{5-56}$$

Multiplying this equation through by V and multiplying and dividing the right-hand side by D yields

$$\mathcal{V}_{j,N} = \frac{V\mathscr{d}_j}{D} = (1 + \mathscr{R})\mathscr{d}_j \tag{5-57}$$

Therefore, knowing the reflex ratio, $\mathcal{V}_{j,N}/\mathscr{d}_j$ is computed as

$$\frac{\mathcal{V}_{j,N}}{\mathscr{d}_j} = 1 + \mathscr{R} \tag{5-58}$$

Inserting this value into Eq. (5-55) for $n = N$ yields $\ell_{j,N}/\mathscr{d}_j$. Inserting this value into Eq. (5-52) for $n = N - 1$ yields $\mathcal{V}_{j,N-1}/\mathscr{d}_j$. This value is then inserted into Eq. (5-55) for $n = N - 1$ to give $\ell_{j,N-1}/\mathscr{d}_j$. This value is then inserted into Eq. (5-52) for $n = N - 2$ to give $\mathcal{V}_{j,N-2}/\mathscr{d}_j$. This procedure is continued, using Eqs. (5-55) and (5-52) alternately, to yield all the $\mathcal{V}/\mathscr{d}$ and $\ell/\mathscr{d}$ values in the enriching section right down to and including $\ell_{j,f+1}/\mathscr{d}_j$ and $\mathcal{V}_{j,f}/\mathscr{d}_j$.

The calculation down the column, just described, yields a value of $\mathcal{V}_{j,f}/\mathscr{d}_j$. The calculation up the column from the bottom, described previously, yields a value of $\ell_{j,f}/\mathcal{L}_j$. These two values are used, together with the equilibrium relationship on the feed plate, to yield a value for the ratio $\mathcal{L}_j/\mathscr{d}_j$. The equilibrium relationship for the feed plate is

$$y_{j,f} = K_{j,f}x_{j,f} \tag{5-59}$$

Multiplying this equation through by V and multiplying and dividing the right-hand side by L' yields

$$\mathcal{V}_{j,f} = \frac{VK_{j,f}}{L'}\ell_{j,f} \tag{5-60}$$

Multiplying and dividing by $\mathscr{d}_j$ and $\mathcal{L}_j$ and rearranging yields

$$\frac{\mathcal{L}_j}{\mathscr{d}_j} = \frac{A_{j,f}(\mathcal{V}_{j,f}/\mathscr{d}_j)}{\ell_{j,f}/\mathcal{L}_j} \tag{5-61}$$

where the absorption factor for component j on the feed plate is defined as

$$A_{j,f} \equiv \frac{L'}{VK_{j,f}} \tag{5-62}$$

Note that this definition is different from that given in Eq. (5-54) for the plates above the feed plate in that it employs the liquid rate in the stripping section and the vapor rate in the enriching section. Using Eq. (5-61) with the value of $\mathcal{V}_{j,f}/d_j$ obtained from the calculation down the column and the value of $\ell_{j,f}/b_j$ obtained from the calculation up the column, a value of b_j/d_j is computed for component j. This entire calculational procedure is employed for each component separately to obtain b_j/d_j for each component in the multicomponent system.

Once the value of b_j/d_j is obtained for each component, the value of b_j and the value of d_j can be obtained from an overall balance around the entire column for component j:

$$b_j + d_j = Fz_{j,F} \tag{5-63}$$

which may be rearranged to give

$$d_j = \frac{Fz_{j,F}}{1 + b_j/d_j} \tag{5-64}$$

Substituting the value of b_j/d_j together with the feed flow rate and the mole fraction of component j in the feed stream into Eq. (5-64) yields a value for d_j. This value is then substituted into Eq. (5-65) to yield a value for b_j:

$$b_j = d_j \frac{b_j}{d_j} \tag{5-65}$$

This equation is to be preferred over the more obvious equation

$$b_j = Fz_{j,F} - d_j \tag{5-66}$$

which would suffer from inaccuracies because of taking the small difference between numbers which are quite close to each other when applied to the light components. Therefore, Eq. (5-65) should be employed, using the values of d_j and b_j/d_j obtained previously.

Multiplying these computed values of b_j and d_j by the values of $\mathcal{V}_{j,n}/d_j$, $\ell_{j,n}/d_j$, $\mathcal{V}_{j,m}/b_j$, and $\ell_{j,m}/b_j$, component flow rates in the liquid and vapor streams can be computed for all components for all plates within the column.

If the values of b_j for all components are added, they will generally not add up to the specified bottom product rate, B. Likewise, if the component flow rates in the liquid-overflow stream from some plate in the stripping section, $\ell_{j,m}$, are added for all components, the sum will generally not equal the specified liquid-overflow rate in the stripping section, L', unless of course the temperature distribution assumed at the start of the calculation was indeed the correct one.

In the original Thiele–Geddes method, the liquid-phase composition on a plate was calculated by dividing the component flow in the liquid stream from that plate by the total of such component flows for all components:

$$x_{j,p} = \frac{\ell_{j,p}}{\sum \ell_{k,p}} \tag{5-67}$$

In this equation plate p can be any plate in the column, in the stripping section, or in the enriching section. The symbol $\sum$ represents a summation on the index k for all components. Similarly, the distillate and bottom product compositions were calculated as

$$x_{j,D} = \frac{\mathscr{d}_j}{\sum \mathscr{d}_k} \tag{5-68}$$

$$x_{j,B} = \frac{\mathscr{b}_j}{\sum \mathscr{b}_k} \tag{5-69}$$

With these newly computed liquid compositions on every plate, the bubble-point temperature for the liquid on every plate was calculated, and a new temperature distribution throughout the column thus determined. This new temperature distribution then provided the starting point for repeating the entire cycle of calculations. Thus a *direct substitution iteration procedure* was used with the whole cycle of calculations being repeated until convergence to the correct solution was obtained.

As an alternative means of obtaining a new temperature distribution throughout the column, dew-point temperatures could be used instead of bubble-point temperatures. The vapor-phase composition for any plate within the column could be calculated as

$$y_{j,p} = \frac{\mathscr{V}_{j,p}}{\sum \mathscr{V}_{k,p}} \tag{5-70}$$

Once the composition of the vapor stream leaving a plate had been determined, the dew-point temperature for that vapor could be used as the next estimate for the temperature on that plate.

5-9. Theta method of convergence

Lyster, Sullivan, Billingsley, and Holland [7, 5] developed a modification of the Thiele–Geddes method which very substantially increases the rate of convergence to the correct solution. These authors have named the method the *theta method of convergence*. The method is based upon the realization that when the guess for the temperature distribution is in error, the values of $\mathscr{b}_j/\mathscr{d}_j$ for the various components will generally all be too high or they will all be too low. In the theta method, these values of $\mathscr{b}_j/\mathscr{d}_j$ are all corrected by the same factor, θ, chosen so that the component flow rates in the distillate and bottom product streams will add to the specified distillate and bottom product flow rates. The calculation of the liquid or vapor compositions throughout the column and of a revised estimate of the temperature distribution throughout the column then proceeds as in the normal Thiele–Geddes procedure. The steps in the theta method are outlined below.

The calculation from the bottom of the column to the feed plate to yield values of $\ell_{j,m}/\mathscr{b}_j$ and $\mathscr{V}_{j,m}/\mathscr{b}_j$ is carried out as in the normal Thiele–

Equation (5-73) then yields θ_{i+1}, the new value of θ for the $(i + 1)$st iteration.

The function $\mathscr{F}(\theta)$ is defined as

$$\mathscr{F}(\theta) \equiv \sum \left[\frac{Fz_{k,F}}{1 + \theta(\ell_k/d_k)}\right] - D \tag{5-74}$$

Differentiating this function yields

$$\mathscr{F}'(\theta) = -\sum \left\{\frac{Fz_{k,F}(\ell_k/d_k)}{[1 + \theta(\ell_k/d_k)]^2}\right\} \tag{5-75}$$

Therefore, the Newton iteration procedure for finding the value of θ that satisfies Eq. (5-72) is as follows. For the value of θ at the ith iteration, namely θ_i, $\mathscr{F}(\theta_i)$ is calculated according to Eq. (5-74). If $\mathscr{F}(\theta_i)$ is suitably close to zero, then θ_i is the desired value of θ. In making this assessment the departure of $\mathscr{F}(\theta_i)$ from zero should be assessed as a percentage of D. For example, when using a digital computer and carrying approximately seven decimal significant figures, a reasonable criterion of convergence would be that $\mathscr{F}(\theta_i)/D$ should be smaller in absolute magnitude than, for example, 10^{-6}. If, on the other hand, $\mathscr{F}(\theta_i)$ is not suitably close to zero, a new value of θ, namely θ_{i+1}, is sought. First, $\mathscr{F}'(\theta_i)$ is calculated at the value θ_i by means of Eq. (5-75). Next, the new value, θ_{i+1}, is calculated from Eq. (5-73). At this point the cycle is repeated until some value of θ is found for which $\mathscr{F}(\theta)$ is indeed suitably close to zero.

This procedure will converge rapidly without any problems provided that no negative values of θ are ever employed. Inspection of Eq. (5-74) shows that a graph of $\mathscr{F}(\theta)$ versus θ for *positive* values of θ appears qualitatively like that shown in Fig. 5-5. For $\theta = 0$, $\mathscr{F}(\theta)$ equals B, the specified bottom product flow rate. For θ approaching infinity, $\mathscr{F}(\theta)$ approaches $-D$, the negative of the specified distillate product flow rate. In between, the curve of $\mathscr{F}(\theta)$ versus θ is a monotonically decreasing function. But inspection of Eq. (5-74) also reveals that $\mathscr{F}(\theta)$ approaches infinity at a number of negative values of θ. Careful study of the Newton iteration procedure and its graphical implications in Fig. 5-5 reveals that if the initial choice of θ is zero, the Newton iteration procedure will converge nicely, with the value of θ_i always increasing toward the correct value of θ that satisfies Eq. (5-72). Therefore, it is often recommended that the initial guess for θ be taken as zero. This will avoid any worries about encountering negative values of θ and therefore about convergence problems with the Newton iteration procedure. On the other hand, as convergence is approached in the Thiele–Geddes calculation, the value of θ for each cycle will get closer and closer to unity, and some computer time can usually be saved by initiating the Newton iteration procedure with a θ value of unity, rather than zero. But if this is done, a safeguard must be employed to ensure that if Eq. (5-73) ever yields a value of θ_{i+1} that is negative, that this value not be employed, but rather that it be changed to zero before the Newton iteration procedure is continued. This procedure is quite satisfactory.

5-11. Use of relative volatilities

The Thiele–Geddes method, employing the theta method of convergence, was described in the previous sections as if the basic vapor–liquid equilibrium data were available in terms of K values for all components as a function of temperature, such as is presented in Fig. 5-1. On the other hand, as has been illustrated in Example Problems 5-3 and 5-4, the use of relative volatilities can simplify the calculation of bubble-point temperatures and dew-point temperatures. It is often convenient, therefore, to express the equilibrium relationship in terms of relative volatilities of the various components as a function of temperature or as a function of the K value for the reference component. In this latter case, the temperature is not considered directly, but rather the K value for the reference component is an indirect measure of temperature. In the Thiele–Geddes procedure, instead of an initial guess of the temperature distribution within the distillation column, an initial guess is made of the K value for the reference component on every plate. Then the K value for any particular component on any plate is simply obtained by multiplying the K value for the reference component by the α value for the component in question. At the end of the calculation cycle, when a vapor composition or a liquid composition on a plate has been determined, a bubble-point calculation or a dew-point calculation would then result in the K value for the reference component on the plate in question rather than in the temperature directly. When the Thiele–Geddes procedure had converged to the final answer, then the temperature on every plate, corresponding to the K value for the reference component, could be determined if the temperature was in fact of interest. The use of the relative volatility approach is especially convenient when the method of modified latent heats of vaporization is employed in a multicomponent distillation-column simulation.

5-12. Modified latent heat of vaporization method

Just as in a binary distillation column, the assumption of constant molal overflow in a multicomponent distillation column will be in error if the molal latent heats of vaporization of the various components are significantly different. As in binary distillation-column design, a considerable improvement upon the assumption of constant molal overflow can be made by assuming, instead, constant pseudo molal overflow, where the pseudo molecular weights of the various components are taken so that the latent heats of vaporization per pseudo mole will be the same for all components.

It is convenient but by no means necessary to choose the pseudo molecular weight of one component to be equal to its actual molecular weight and then to choose pseudo molecular weights for all the other components so that all components have the same pseudo molal latent heat of vaporization.

When the assumption of constant pseudo molal overflow is employed, the use of relative volatilities in expressing the equilibrium relationships is especially useful because of the invariance of the relative volatility, as explained in Chap. 4. Thus, Eqs. (5-3), (5-10), and (5-15) apply whether true mole fractions or pseudo mole fractions are employed, and the value of α is the same for a given liquid-phase composition. The recommended procedure, therefore, is as follows. Assume that the liquid-phase composition on a given plate has just been computed in the Thiele–Geddes method in terms of pseudo mole fractions. The problem then is to determine a new "temperature" for that plate in order that another cycle of the calculation can be performed. First, the pseudo mole fractions in the liquid phase are converted to true mole fractions. Then, a bubble-point calculation is performed in a manner similar to that in Example Problem 5-3. The "temperature" could easily be replaced by the K value for the reference component. When this calculation is complete, accurate relative volatility values for all components can be determined at the bubble-point temperature or at the corresponding K value for the reference component. This K value for the reference component will be the true K value, that is, the ratio of the true mole fraction of the reference component in the vapor phase to the true mole fraction of the reference component in the liquid phase at equilibrium. Once accurate relative volatility values are obtained, the calculation can then proceed completely in terms of pseudo molal quantities. Using the relative volatilities and the liquid-phase pseudo mole fractions, a pseudo K value for the reference component can be calculated with Eq. (5-9), employing pseudo mole fractions. This pseudo K value for the reference component would then be the ratio of the vapor-phase pseudo mole fraction to the liquid-phase pseudo mole fraction for the reference component. Next, the pseudo K value for any component on that plate could be calculated as the product of the pseudo K value for the reference component and the relative volatility for the component in question. Therefore, pseudo K values for all components on that plate could be readily determined for the next cycle of the Thiele–Geddes calculation.

If, instead of liquid-phase compositions, vapor-phase compositions were employed, the procedure would be similar. Assume that the vapor-phase composition, in terms of pseudo mole fractions, is known on any plate. These vapor-phase pseudo mole fractions can be converted to true vapor-phase mole fractions. Then a dew-point calculation can be carried out according to the method of Example Problem 5-4. In this calculation the K value for the reference component could conveniently be substituted for temperature. The calculation would yield the true K value for the reference component, which is a measure of plate temperature, and also accurate relative volatility values for all components at that temperature. These relative volatility values would then be employed in Eq. (5-14), using vapor-phase pseudo mole fractions, to yield the pseudo K value for the reference component. Multiplying the pseudo K value for the reference component by the relative volatility for any component would yield the pseudo K value for that component on the plate in question.

Once the pseudo K values for all components on all plates were determined, the Thiele–Geddes calculation would proceed in terms of pseudo molal flow rates and pseudo mole fractions until new liquid or vapor compositions on all plates were determined at the end of the cycle. Then these values would be converted to real mole fractions to perform the bubble-point or the dew-point calculation, and the whole cycle would be repeated again.

The assumption of constant pseudo molal overflow in a multicomponent distillation column will be in error, just as in a binary distillation column, if the heat of mixing in the liquid phase is large or if sensible heat effects in the liquid phase are large. Generally speaking, the difference between the boiling points of the least volatile component and the most volatile component will be much greater in a multicomponent column than in a typical binary distillation column; therefore, sensible heat effects are more likely to be important in the multicomponent case. Indeed, the boiling points of some of the components at the column pressure may be far outside of the range of temperatures existing within the distillation column. In such cases it may be difficult to select a value of the latent heat of vaporization for each component in order to assign pseudo molecular weights such that all components will have the same pseudo molal latent heat. It will generally be more accurate to employ latent heats of vaporization at some average temperature, perhaps the average of the reboiler and top-plate temperatures, rather than to use the latent heat of each component at its boiling point at the column pressure.

MULTICOMPONENT DISTILLATION-COLUMN PERFORMANCE

It is instructive to examine in some detail the behavior of a typical multicomponent distillation-column simulation. Accordingly, a computer program using the Thiele–Geddes–theta method previously described was used to simulate column performance for the system described in Example Problem 5-6. First, the relative volatilities were assumed to be constant at the values taken in the solution to Example Problem 5-6. Next, variations in the relative volatilities were taken into account by fitting the data in Fig. 5-2 to broken-line relationships between the relative volatilities and the K value for isobutane, which was used to represent temperature.

5-13. Design curve

For each of several values of N chosen, the value of $\mathscr{R}$ was varied until, by trial and error and cross plotting, the product purity specifications

$$x_{4,D} = x_{3,B} = 0.005$$

were met or closely approached. In this discussion the component subscripts refer to the components in order of volatility; thus 3 refers to n-butane and 4 refers to isopentane. In addition to varying $\mathscr{R}$, the feed-plate location had to be adjusted by trial and error so that the product purities were maximized at each value of $\mathscr{R}$.

The results of these calculations are summarized in Fig. 5-6, the design curve of N versus $\mathscr{R}$. Shown for comparison are the approximate design curve and the limiting values $N_{\min}$ and $\mathscr{R}_{\min}$ from the solution to Example Problem 5-6. It is seen that the approximate solution of Example Problem 5-6 agrees closely with the Thiele–Geddes–theta method of simulation using constant relative volatilities. On the other hand, taking account of the variations in the relative volatilities shifts the design curve appreciably, as much as a 10 per cent shift in $\mathscr{R}$.

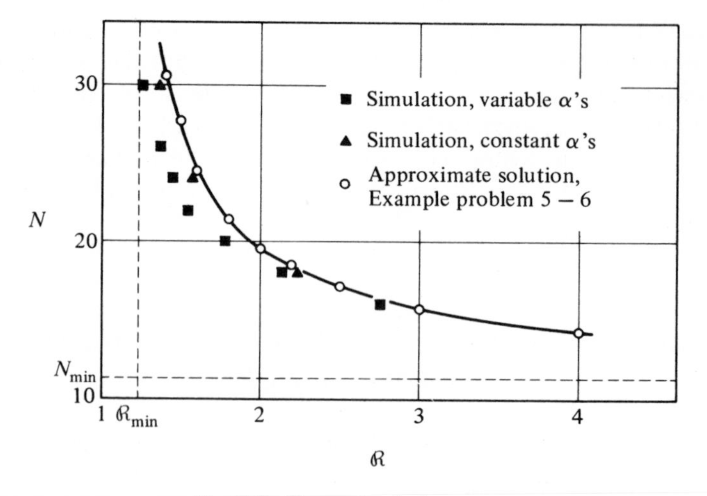

Fig. 5-6. Computer design curve.

5-14. Composition profiles

Considerable insight into the behavior of a multicomponent distillation system can be gained by examining the composition profiles throughout the column for various reflux ratios. The profiles presented here were taken from the simulation with constant relative volatilities, but the variations in the relative volatilities do not affect the conclusions to be drawn.

Figure 5-7 presents the profiles of liquid-phase mole fractions for $N = 30$, $\mathscr{R} = 1.354$. In the bottom product, $x_4 = 0.359$ and $x_6 = 0.3635$, as given in the solution to Example Problem 5-6. Moving up the column toward the feed plate, x_4 is seen to rise to a maximum value of 0.53 and then to fall off to 0.24 at the feed plate. In contrast, x_6 falls rapidly at first and then levels out, becoming almost asymptotic at a value of 0.145 at the feed plate. The behavior of x_5 is intermediate between that of x_4 and that of x_6, showing some of the characteristics of each.

The appearance of a maximum in a composition profile is not found in binary distillation, but it is quite common in multicomponent distillation. It can be explained as follows. In the bottom product, x_1, x_2, and x_3 sum to less than 0.0051, and thus the system behaves almost as in the distillation of the ternary system of components 4, 5, and 6. In this ternary system, component 4 is the most volatile and component 6 is the least volatile component, and so it is obvious that x_4 must rise and x_6 must fall with increasing distance up the column. But component 3 is, of course, more volatile than component 4; so x_3 rises much faster than x_4, and when plate 4 is reached, x_3 has risen to 0.062, which is no longer negligible compared to unity. Continuing up the column, as x_3 continues to rise it is clear that x_4, x_5, and x_6 must fall in order that $\sum x_k$ may remain equal to unity on each plate. All the while, component 2 is, of course, more volatile than component 3, although its effect is not felt while x_2 is negligible compared to unity. But by plate 13, x_2 has reached a value of 0.045, and as x_2 continues to rise above plate 13, x_3 can be seen to pass through a maximum and then decrease slightly toward its value at the feed plate.

Concentration maxima for components of intermediate volatility are common in multicomponent distillation columns, and sometimes the maxima are very pronounced. Argon, for example, peaks up to 15 or 20 per cent in the stripping section of an air distillation column, even though it is only about 1 per cent in the feed and in the distillate product and only about 0.1 per cent in the oxygen bottom product. This fact is of importance in argon recovery by means of a *side column*, which removes the argon from the vapor stream withdrawn from the appropriate point in the stripping section of the air distillation column.

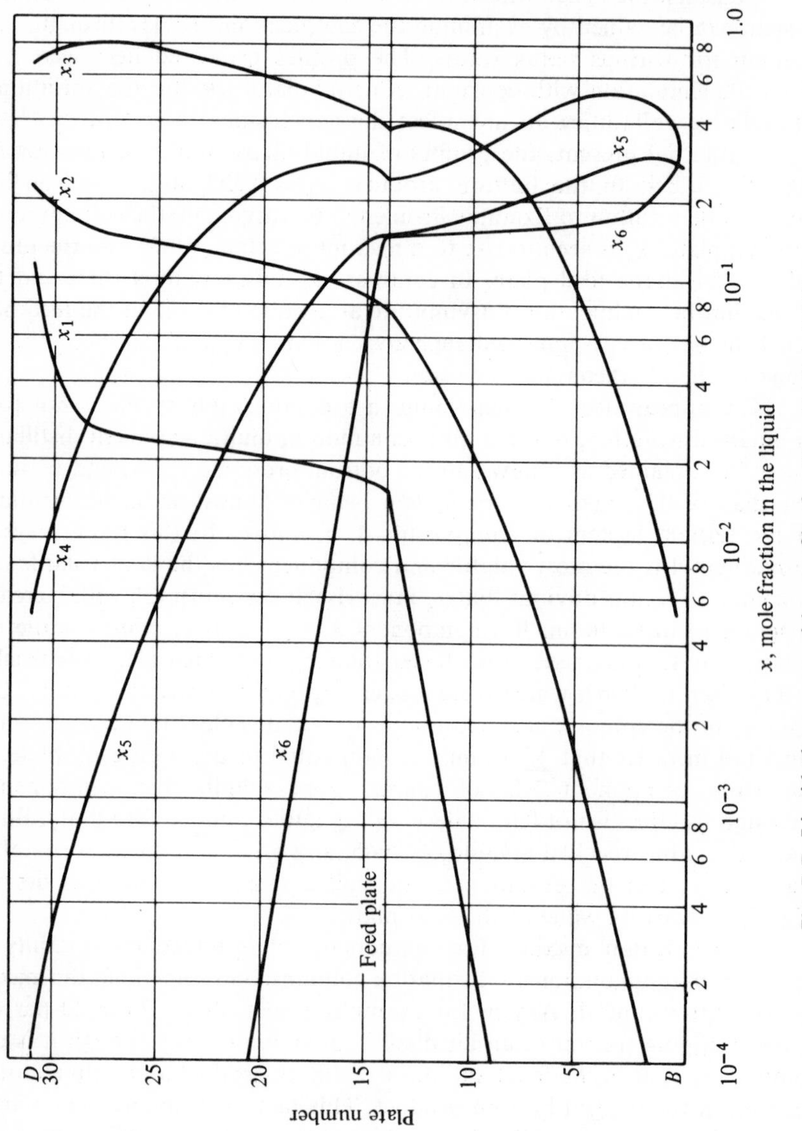

Fig. 5-7. Liquid-phase composition profiles for $N = 30$, $\mathscr{R} = 1.354$.

5-15. Material-balance limits on the compositions

It should be noted in Fig. 5-7 that as x_3 rises from 0.062 on plate 4 to 0.345 on plate 10, x_4 falls by a much greater percentage than does x_6. This is because x_6 is asymptotically approaching a limiting value below which it cannot fall. This limiting value, x_6^0, is given by the expression

$$x_j^0 L' = Bx_{j,B} \tag{5-76}$$

and can be seen to be the value of x_j required to carry $Bx_{j,B}$ moles per unit time of component j from the feed plate to the bottom of the column when any upward flow of component j in the vapor phase is *neglected.* With $\mathscr{R} =$ 1.354, $L' = 160.9$, and Eq. (5-76) yields $x_6^0 = 0.1243$ in this case.

The temperature is high at the bottom of the column, and thus K_6 is high, and x_6 is substantially greater than x_6^0 because the the term $V'y_6$ is by no means negligible in the material balance [see Eq. (4-10)]. The temperature decreases and K_6 falls off with increasing distance up the column, and $V'y_6$ becomes relatively insignificant compared to $Bx_{6,B}$, and x_6 asymptotically approaches $x_6^0 = 0.1243$. It is seen in Fig. 5-7 that at the feed plate x_6 is only 17 per cent greater than x_6^0. Above the feed, however, the material-balance restriction implied by Eq. (5-76) is removed; there is no required flow rate of component 6 in the enriching section of the column, and x_6 falls off very rapidly above the feed plate.

Equation (5-76) can also be used to compute $x_5^0 = 0.093$ and $x_4^0 = 0.123$. Figure 5-7 reveals that at the feed plate x_5 is 61 per cent greater than x_5^0, and x_4 is 95 per cent greater than x_4^0. Since components 4 and 5 are more volatile than component 6, x_4 and x_5 do not approach their limiting values as closely as x_6 does; therefore, x_4 and x_5 are changing by greater percentages than x_6 as the temperature falls off as the feed plate is approached from beneath.

Figure 5-8 presents the vapor-phase composition profiles throughout the column. They can be seen to be qualitatively similar to the liquid-phase composition profiles, but the y curves for the heavy components are not nearly so flat as the x curves as the feed plate is approached from beneath. The reason is that the temperature is decreasing and therefore the K values of all components are falling. Thus, while x_6 is varying but little, y_6 will vary because K_6 is varying. The term $V'y_6$ in the material balance is about one fifth as great as the term $Bx_{6,B}$; therefore, x_6 is about 17 per cent greater than x_6^0, and the percentage variation in x_6 is about one sixth as great as the percentage variation in y_6.

For a light component above the feed plate, there is a limiting value of y required to carry the component up to the top of the column, even when any downward flow of that component in the liquid phase is neglected:

$$Vy_j^0 = Dx_{j,D} \tag{5-77}$$

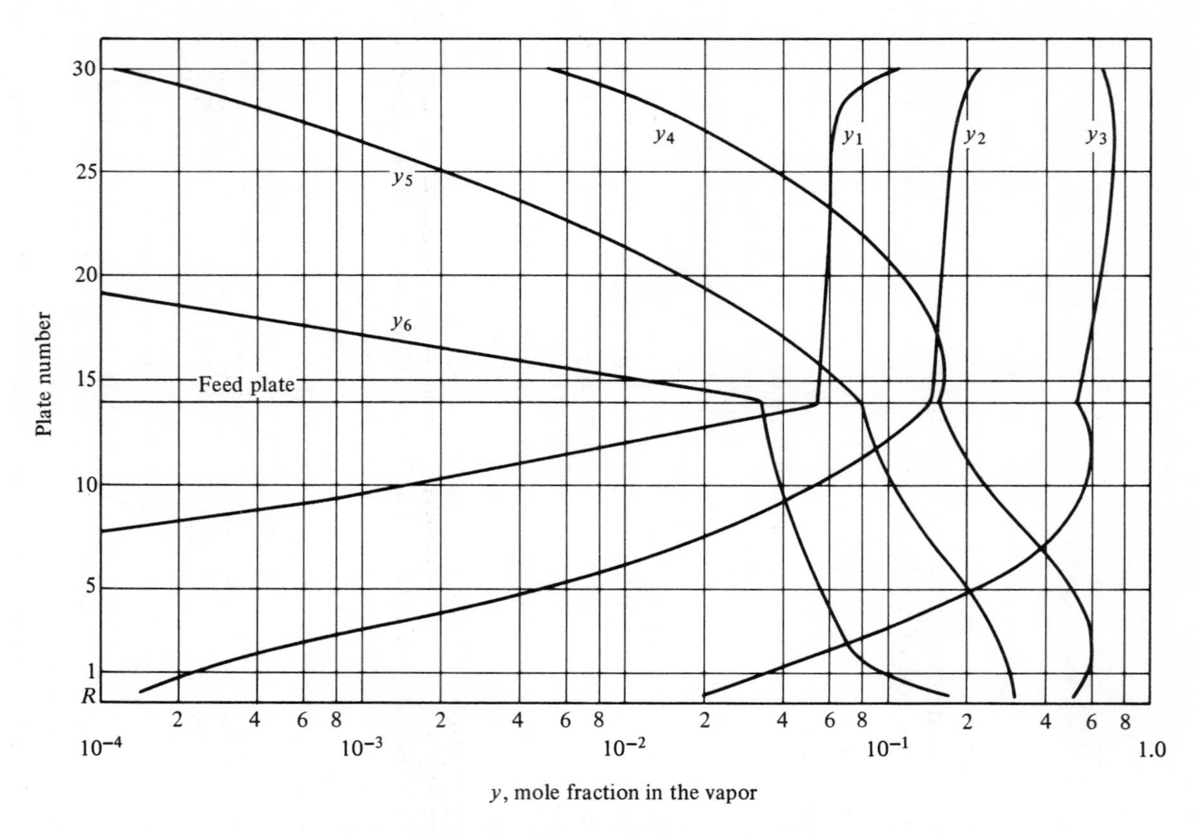

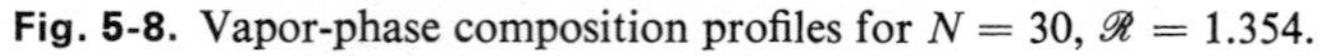

Fig. 5-8. Vapor-phase composition profiles for $N = 30$, $\mathscr{R} = 1.354$.

For the present case, with $\mathscr{R} = 1.354$, $V = 105.8$, and $y_1^0 = 0.0472$. At the top of the column the term Lx_1 is comparable to the term $Dx_{1,D}$ in the enriching-section material balance [see Eq. (4-12)], and therefore y_1 is about twice as great as y_1^0. Moving down the column toward the feed plate, the temperature rises and therefore K_1 rises, making x_1/y_1 smaller. The term Lx_1 soon becomes quite small compared to $Dx_{1,D}$, and y_1 approaches y_1^0 asymptotically. It is seen in Fig. 5-8 that y_1 decreases rapidly from 0.1113 at the top plate and becomes asymptotic, reaching a value of 0.054 at the feed plate. This value is 14 per cent greater than y_1^0, indicating that Lx_1 is only one sixth as great as $Dx_{1,D}$ at this point. Even though the y_1 profile becomes quite flat, the x_1 profile continues to change by a much larger percentage, reflecting the increase in K_1. Of course, below the feed plate the material-balance restriction implied by Eq. (5-77) is removed, and y_1 falls off quite rapidly below the feed plate. The behavior of y_2 and y_3 is qualitatively similar to that of y_1, but y_2 and y_3 are not as nearly asymptotic because the Lx_j terms are not so negligible in these cases.

Figure 5-8 shows that y_3 first starts to rise and then passes through a maximum and then decreases with increasing distance down the column. This behavior is similar to that of x_4 in the bottom of the column. It reflects the fact that component 3 is the heavy component in the apparently ternary system at the top of the column, but that it must ultimately yield to the even heavier component 4 when that component builds up to an appreciable concentration.

5-16. Approach to minimum reflux

Figures 5-9 and 5-10 present the composition profiles for $N = 40$, $\mathscr{R} = 1.2743$, and Figs. 5-11 and 5-12 present the profiles for $N = 80$, $\mathscr{R} = 1.240$. This latter case is quite close to the minimum reflux ratio for the product specifications employed. The general features seen in Fig. 5-7 are also seen in Figs. 5-9 and 5-11. In addition, however, another feature appears and becomes more pronounced as the minimum reflux ratio is approached. This is a zone in the middle of the stripping section in which x_3, x_4, x_5, and x_6 hardly change from plate to plate. This is called a lower *pinch zone*, and it is analogous to the pinch between the lower operating line and the equilibrium curve in a binary distillation column at minimum reflux (see Chap. 4). In contrast to the binary case, the pinch zone in Fig. 5-11 occurs not at the feed plate but below it. This is so because right at the feed plate x_1 and x_2 are not negligible, and several plates beneath the feed plate are required in order for x_1 and x_2 to fall to negligible values. Even then, x_1 and x_2 continue to fall off by large percentages, but their values are so small that they do not affect the values of x_3, x_4, x_5, and x_6. The temperature also changes hardly at all from plate to plate, and so y_3, y_4, y_5, and y_6 are also almost

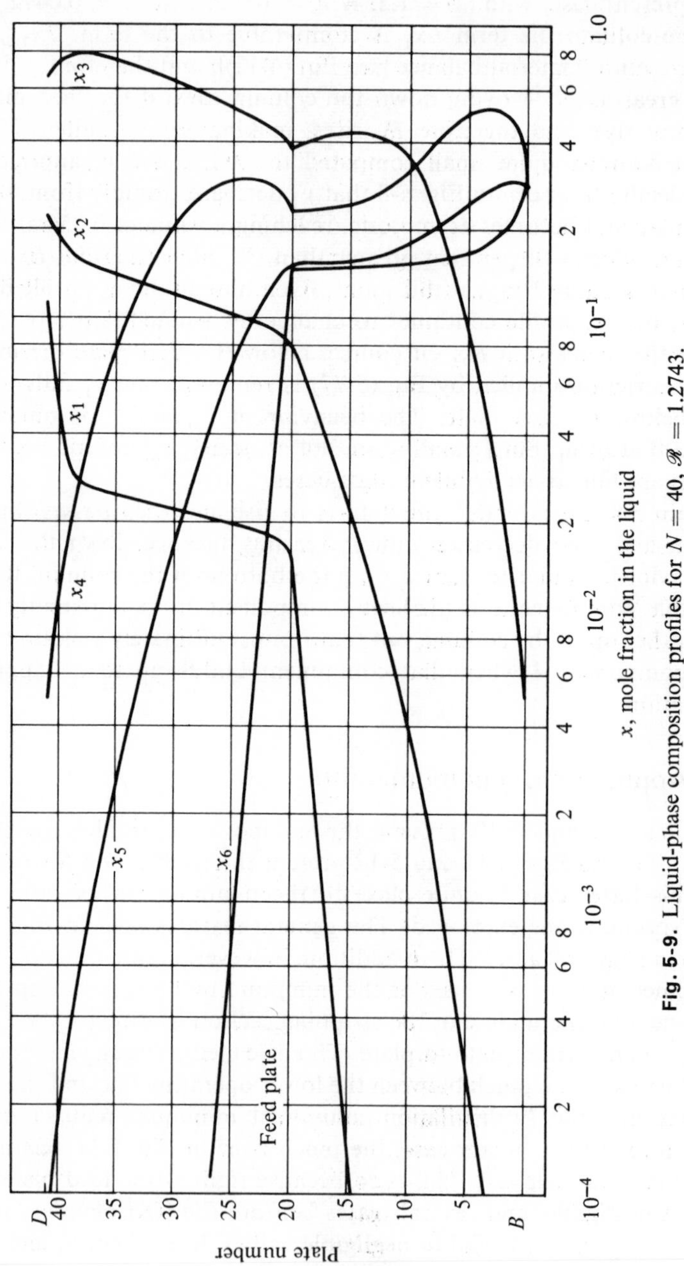

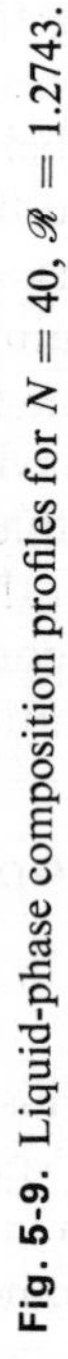

Fig. 5-9. Liquid-phase composition profiles for $N = 40$, $\mathscr{R} = 1.2743$.

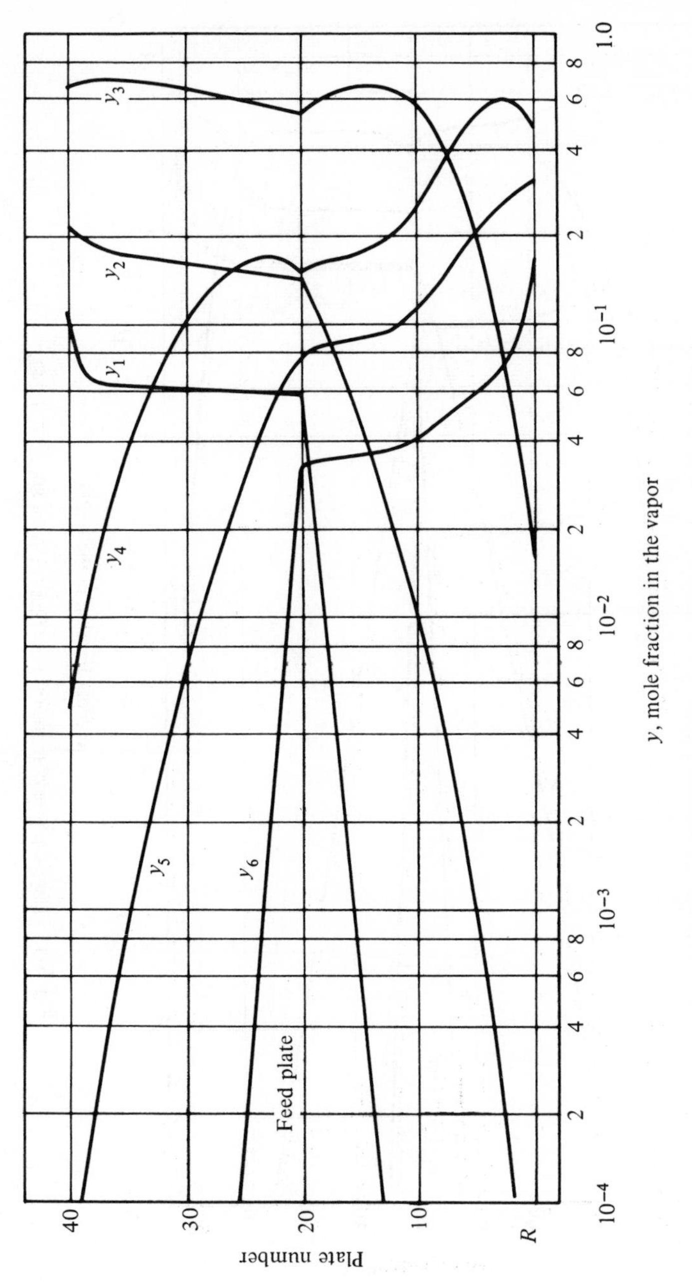

Fig. 5-10. Vapor-phase composition profiles for $N = 40$, $\mathscr{R} = 1.2743$.

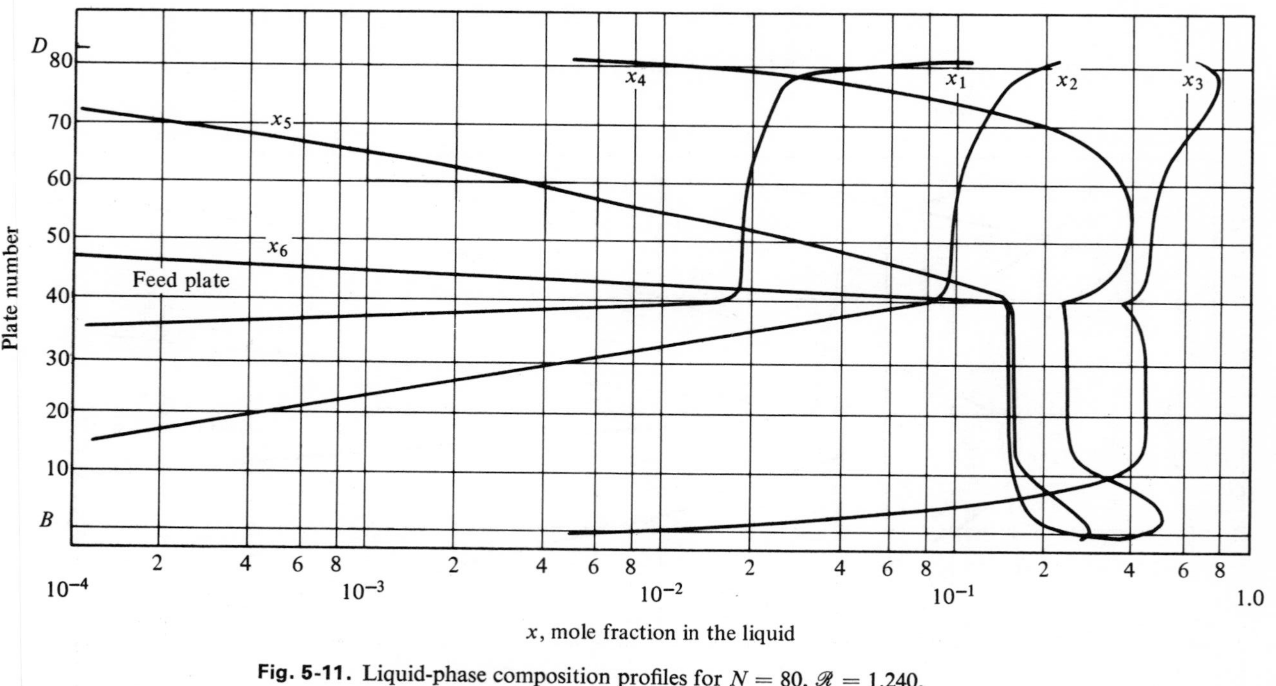

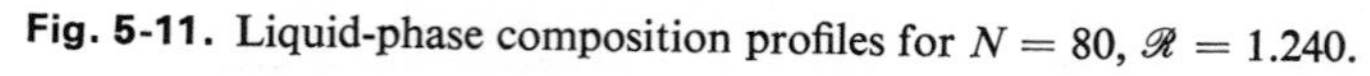

Fig. 5-11. Liquid-phase composition profiles for $N = 80$, $\mathscr{R} = 1.240$.

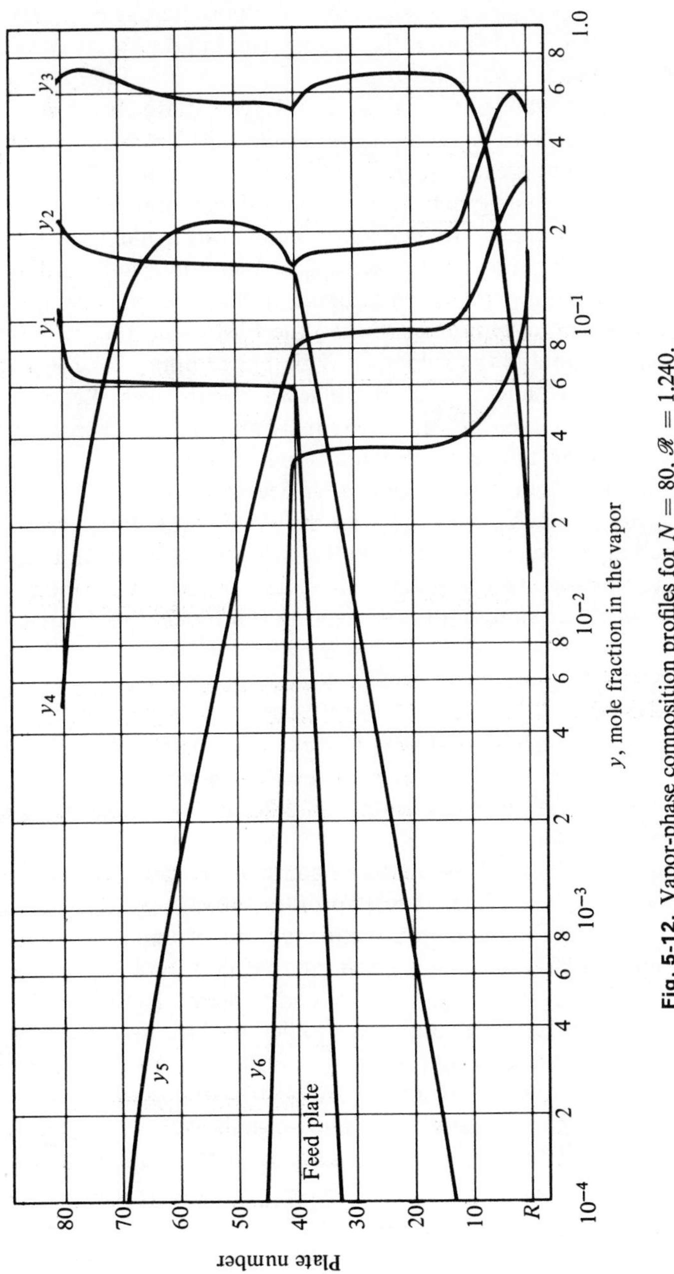

Fig. 5-12. Vapor-phase composition profiles for $N = 80$, $\mathscr{R} = 1.240$.

constant in the lower pinch zone. Thus the pinch zone is essentially free of components 1 and 2 and is essentially isothermal and of constant liquid and vapor compositions. Above this zone, x_3, x_4, x_5, and x_6 fall off as x_1 and x_2 increase rapidly toward their values at the feed plate, but the percentage decrease is much greater for component 3 than it is for component 6 because x_6 is only 18 per cent greater than x_6^0, while x_3 is 250 times x_3^0.

Similarly, an upper pinch zone appears in the middle of the enriching section in which x_5 and x_6 are negligibly small and in which the temperature and the liquid and vapor mole fractions of components 1, 2, 3, and 4 are hardly changing from plate to plate. Beneath this zone, y_1, y_2, y_3, and y_4 fall off as y_5 and y_6 increase rapidly toward the feed plate, but y_4 falls off by a much larger percentage than does y_1. This upper pinch zone is not seen as clearly in Figs. 5-11 and 5-12 as is the lower pinch zone. Larger values of N and reflux ratios even nearer to the minimum value are required in this case to develop the upper pinch zone more clearly.

As N is increased further and $\mathscr{R}$ is decreased toward $\mathscr{R}_{\min}$, after the upper and lower pinch zones are well developed, the composition profiles remain unchanged except that the pinch zones contain more and more plates. That is, the profiles stop changing above and below each pinch zone, and all additional plates simply go into making the pinch zones larger.

5-17. High reflux ratios

Figures 5-13 and 5-14 present composition profiles for $N = 18$, $\mathscr{R} = 2.238$. At this higher reflux ratio, x_6 is seen to be varying more steeply just beneath the feed plate than it was in Fig. 5-7, and a calculation reveals that x_6 is 22 per cent greater than x_6^0.

Figure 5-15 shows the liquid-phase composition profiles for total reflux, $\mathscr{R} = \infty$, $N = N_{\min} = 11.35$. These profiles were not obtained with the simulation computer program but rather by use of equations such as Eqs. (5-19) through (5-28). At $\mathscr{R} = \infty$, the internal flow rates approach infinity, and the feed flow rate is negligible by comparison. Therefore, there is no significance to a feed plate, and the profiles in Fig. 5-15 have no abrupt changes in slope such as those shown in Figs. 5-7 through 5-14. Since L' and V approach infinity, Eqs. (5-76) and (5-77) require that x^0 and y^0 approach zero for all components, and the profiles show no asymptotic behavior such as is shown to varying degrees in Figs. 5-7 through 5-14.

5-18. Retrograde fractionation

Figures 5-16 through 5-19 present semilogarithmic profiles of the *key ratio*, x_3/x_4, throughout the column for reflux ratios of 2.238, 1.354, 1.2743, and 1.240, respectively. The plot for $\mathscr{R} = \infty$ is not shown, but it is clear

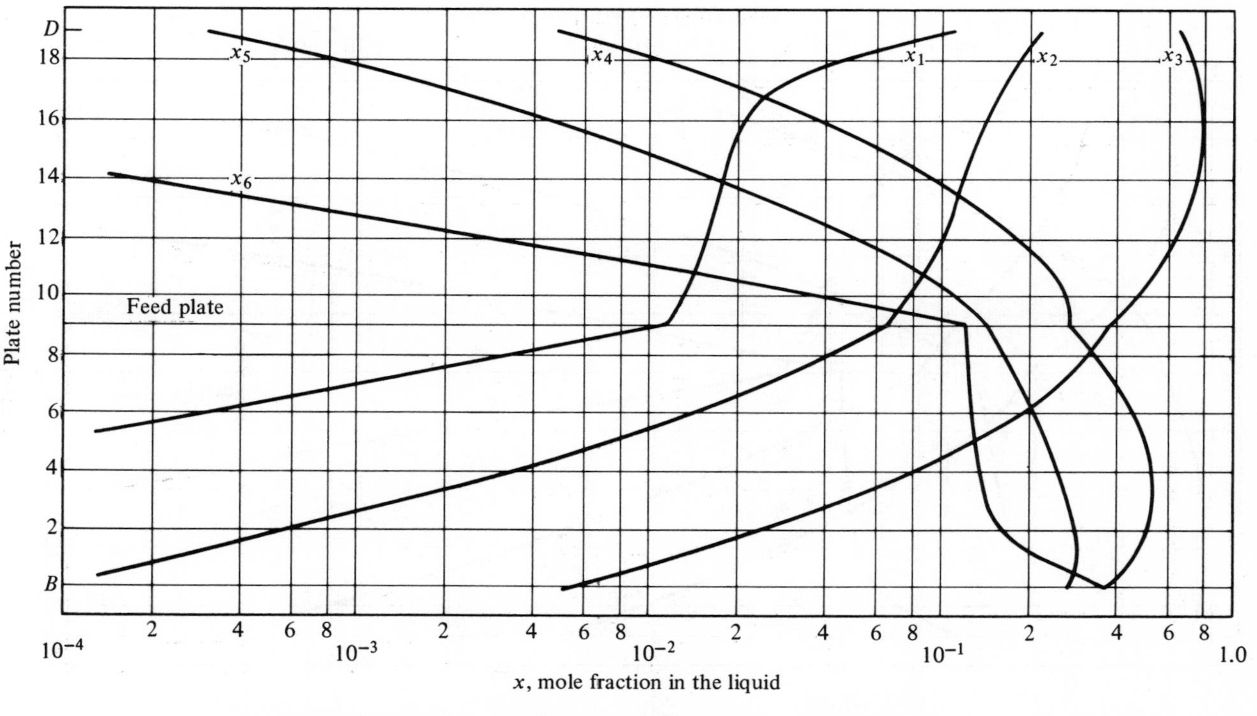

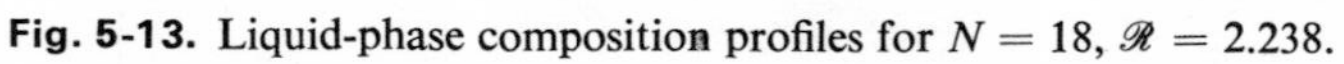

Fig. 5-13. Liquid-phase composition profiles for $N = 18$, $\mathscr{R} = 2.238$.

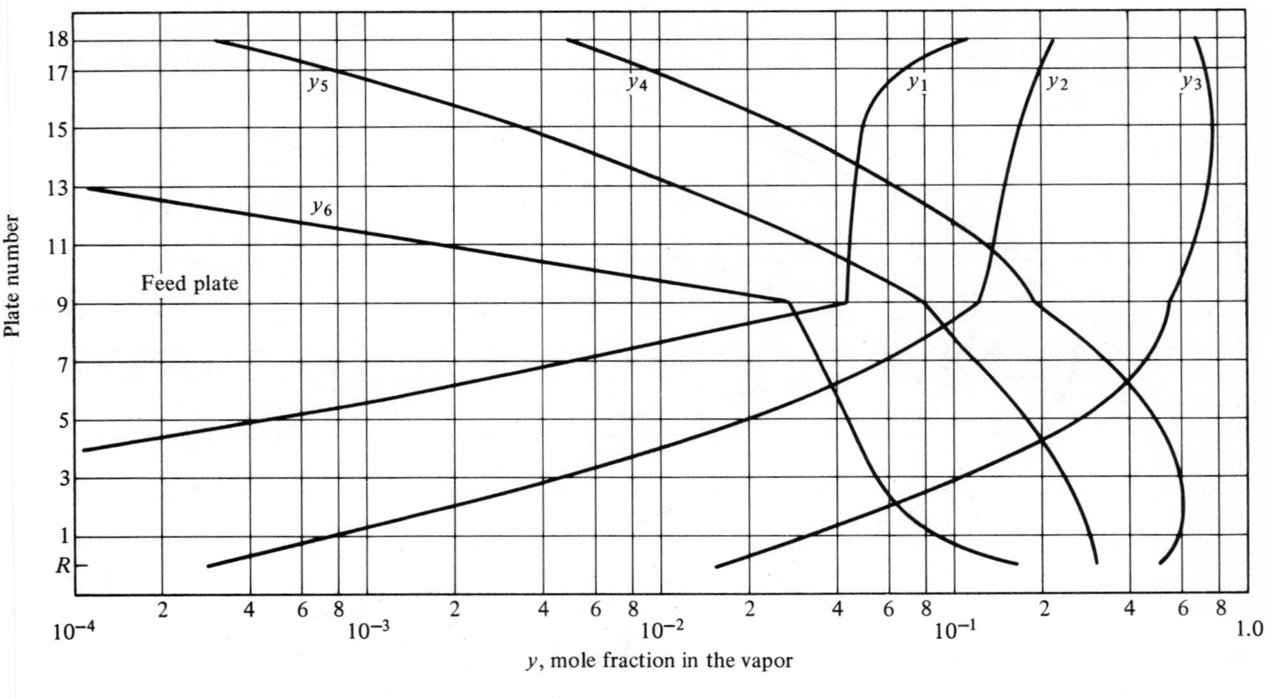

Fig. 5-14. Vapor-phase composition profiles for $N = 18$, $\mathscr{R} = 2.238$.

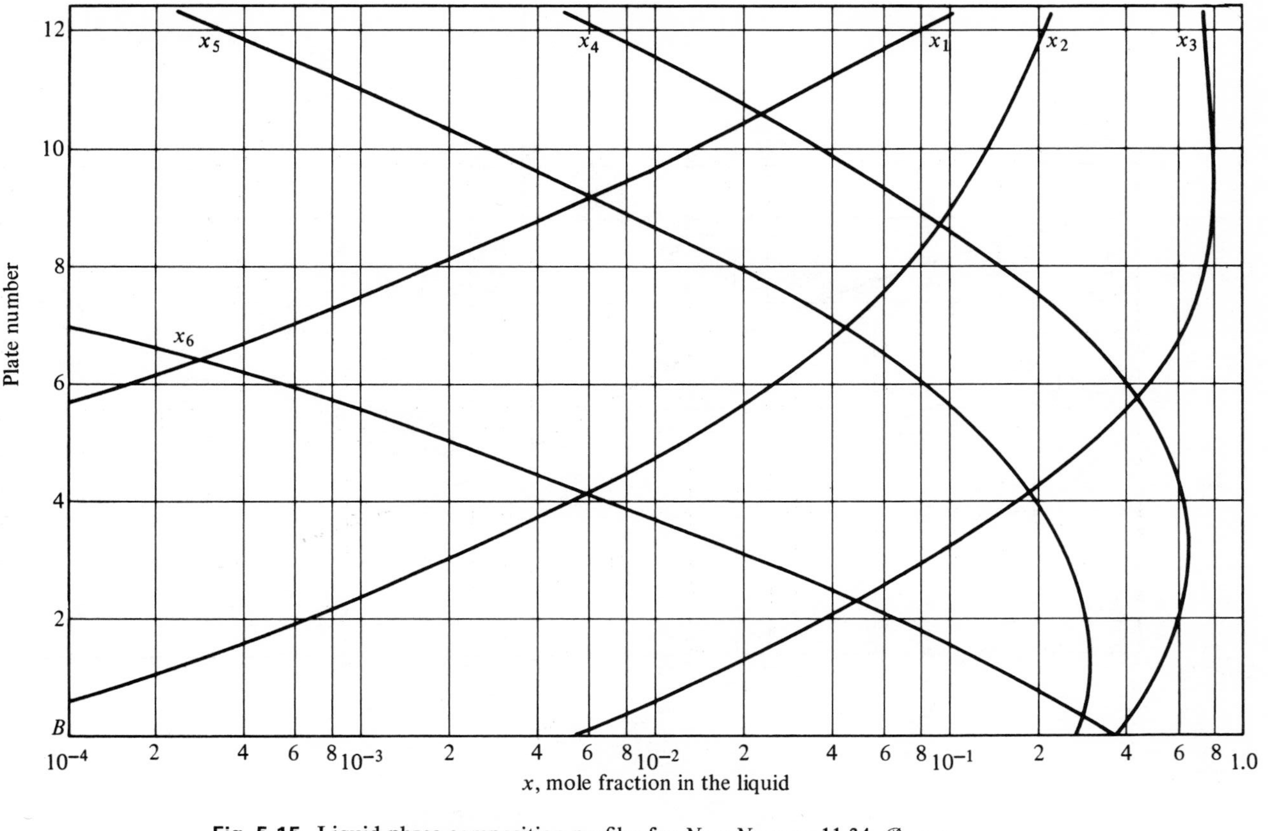

Fig. 5-15. Liquid-phase composition profiles for $N = N_{\text{MIN}} = 11.34$, $\mathscr{R} = \infty$.

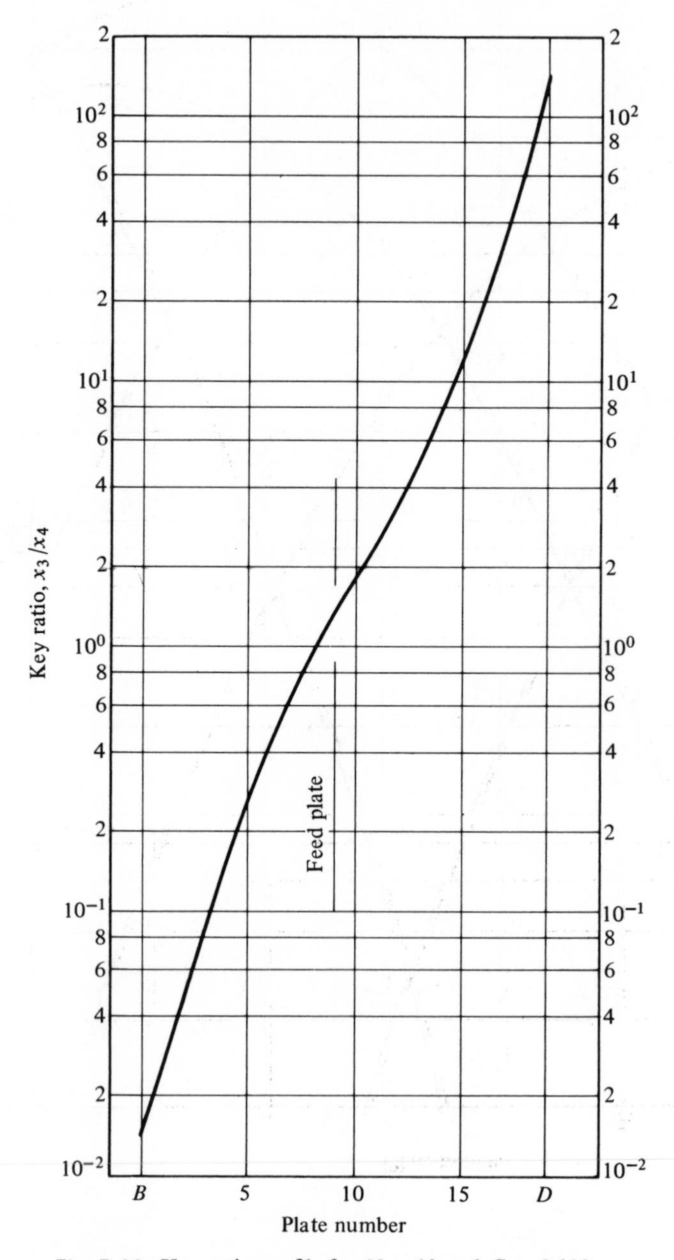

Fig. 5-16. Key-ratio profile for $N = 18$ and $\mathscr{R} = 2.238$.

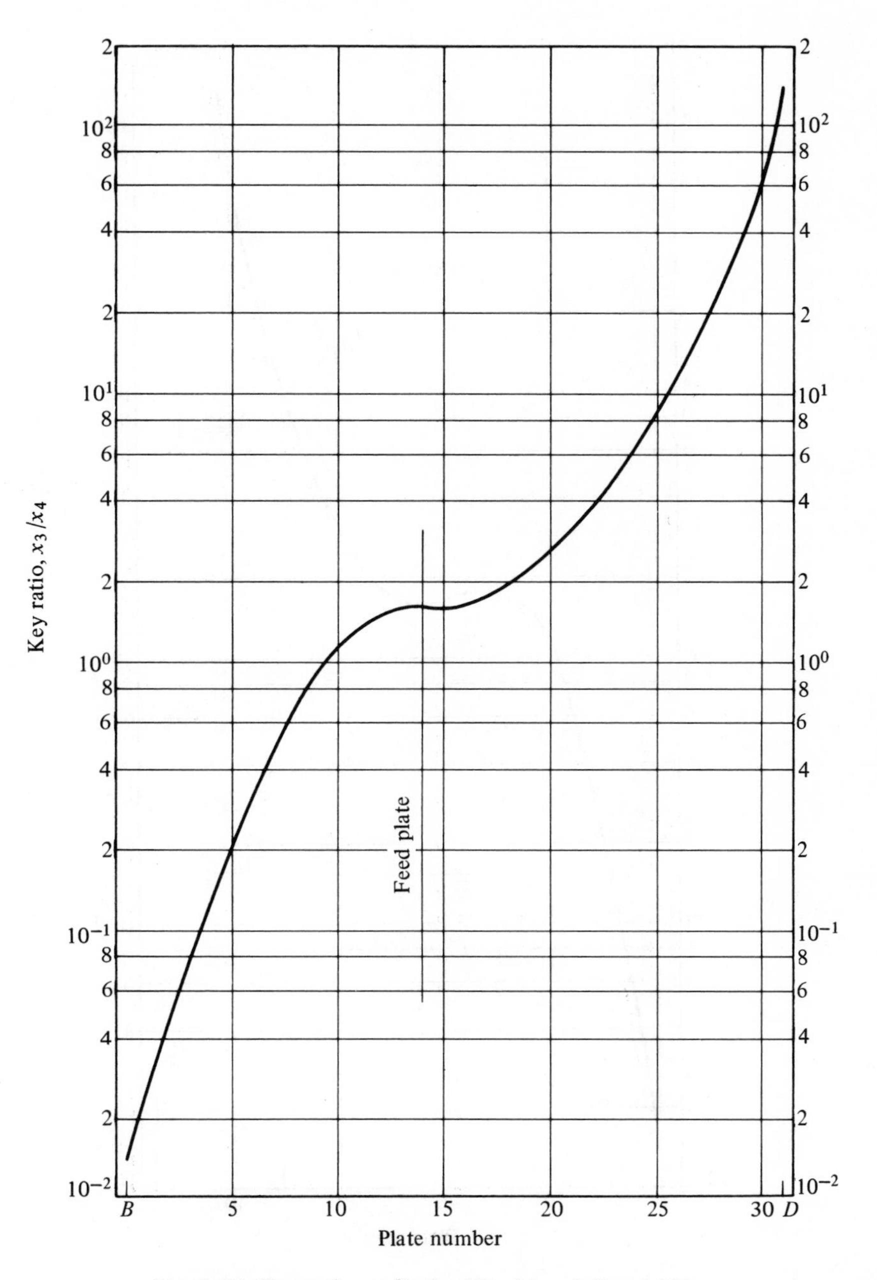

Fig. 5-17. Key-ratio profile for $N = 30$ and $\mathscr{R} = 1.354$.

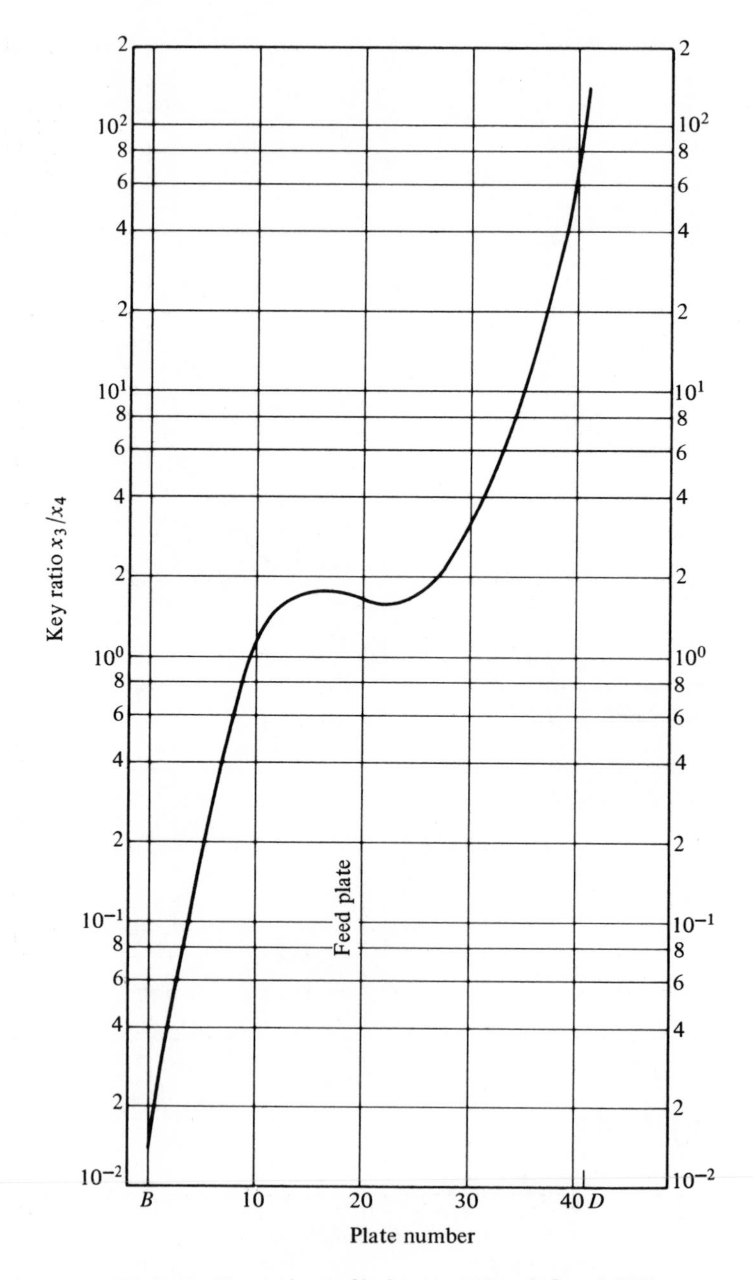

Fig. 5-18. Key-ratio profile for $N = 40$ and $\mathscr{R} = 1.2743$.

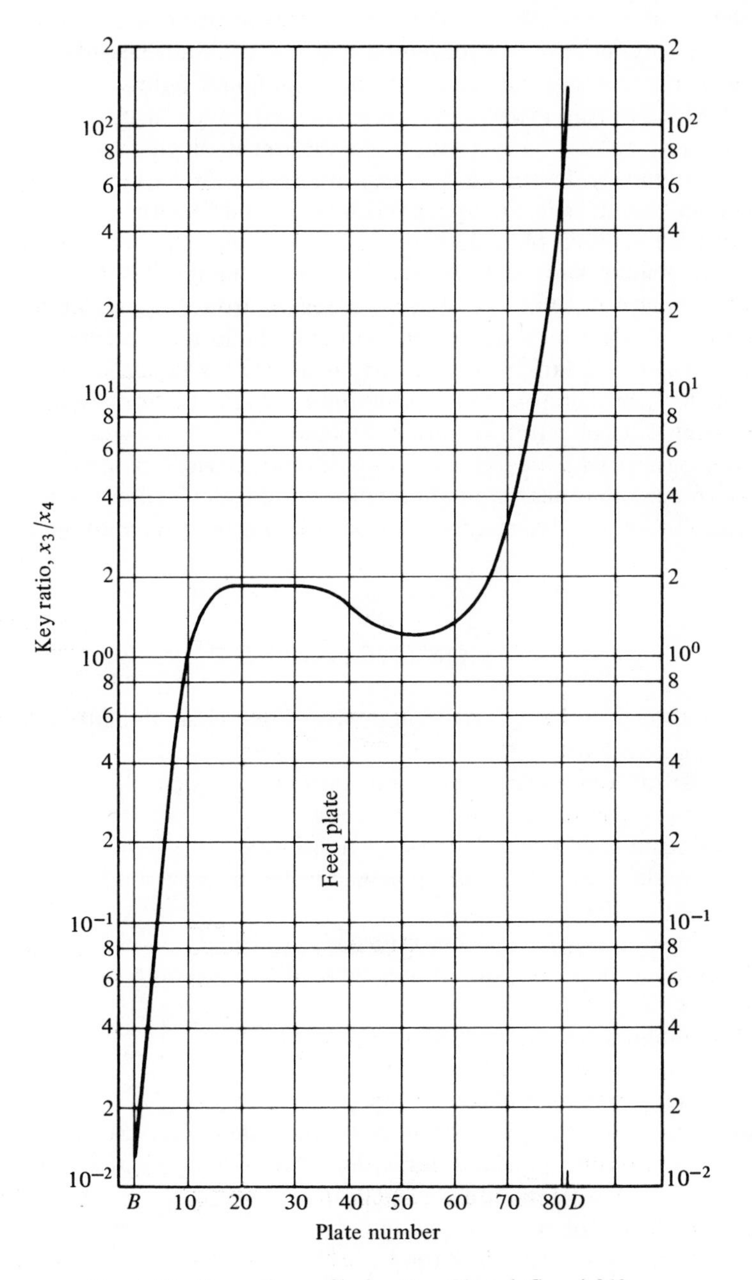

Fig. 5-19. Key-ratio profile for $N = 80$ and $\mathscr{R} = 1.240$.

from Eqs. (5-19) through (5-26) that this profile is linear when plotted on semilogarithmic coordinates. Indeed, even for $\mathscr{R} = 2.238$, shown in Fig. 5-16, the key-ratio profile departs but little from a straight line through the specified end points. But as the reflux ratio is decreased further toward $\mathscr{R}_{\min}$, the key-ratio profile develops a more pronounced inflection point and then increasingly flat maximum and minimum points.

In Fig. 5-19 the lower pinch zone is clearly marked by some 16 plates of essentially constant key ratio, while the upper pinch zone is seen to be not yet completely developed. The key ratio is seen to be higher in the lower pinch zone than it is in the upper pinch zone, and thus there is a *retrograde fractionation* between plate 33 and plate 53 in which the key ratio decreases rather than increasing with distance up the column. This behavior is not found in binary distillation columns. It results from the greater percentage decrease in x_3 than in x_4 as x_1 and x_2 rapidly build up just beneath the feed plate, as explained earlier, and also from the analogous behavior above the feed plate. It is therefore most pronounced when the nonkey components are in high concentration at the feed plate.

Retrograde fractionation is a unique characteristic of multicomponent distillation columns near minimum reflux, and it is important in the Gilliland [3, 11] and Colburn [1] methods for calculating the minimum reflux ratio.

NOMENCLATURE

$A_{j,f}$ = absorption factor for component j on the feed plate, defined by Eq. (5-62)

$A_{j,n}$ = absorption factor for component j on plate n, a plate in the enriching section, defined by Eq. (5-54)

B = bottom product flow rate, moles/time

$\mathscr{b}_j$ = molal flow rate of component j in the bottom product, defined by Eq. (5-34)

D = distillate product flow rate, moles/time

$\mathscr{d}_j$ = molal flow rate of component j in the distillate product, defined by Eq. (5-50)

F = molal flow rate of feed stream, moles/time

$\mathscr{F}(\theta)$ = function of θ defined by Eq. (5-74)

$\mathscr{F}'(\theta)$ = derivative of $\mathscr{F}(\theta)$ with respect to θ; see Eq. (5-75)

$f(\phi)$ = function of ϕ equal to the left-hand side of Eq. (5-30)

K_j = equilibrium constant for component j, $\equiv y_j/x_j$ at equilibrium

$K_{j,p}$ = K_j at the temperature of plate p

K_{ref} = K value for reference component

L = molal flow rate of liquid-overflow stream in enriching section, moles/time

L' = molal flow rate of liquid-overflow stream in stripping section, moles/time

$\ell_{j,p}$ = molal flow rate of component j in the liquid-overflow stream from plate p; see Eqs. (5-32) and (5-48)

N = number of (theoretical) plates in the column; plate number of the top plate

$\mathcal{N}$ = number of equilibrium stages, including theoretical plates and also equilibrium partial reboiler or condenser

P = column pressure, atm

p_j^0 = vapor pressure of component j at the temperature in question, atm

q = liquid fraction in feed, as defined in Chap. 4

$\mathcal{R}$ = reflux ratio $\equiv L/D$

$S_{j,m}$ = stripping factor for component j on plate m, a plate in the stripping section, defined by Eq. (5-41)

$S_{j,R}$ = stripping factor for component j in the reboiler, defined by Eq. (5-37)

$\mathcal{S}_{j,k}$ = separating power, defined by Eq. (5-29)

V = molal vapor flow rate in enriching section, moles/time

V' = molal vapor flow rate in stripping section, moles/time

$\mathcal{V}_{j,p}$ = molal flow rate of component j in the vapor stream leaving plate p; see Eqs. (5-33) and (5-49)

x_j = liquid-phase mole fraction of component j

$x_{j,p}$ = mole fraction of component j in liquid overflow from plate p

y_j = vapor-phase mole fraction of component j

$y_{j,p}$ = mole fraction of component j in vapor stream leaving plate p

$z_{j,F}$ = mole fraction of component j in the feed stream

GREEK LETTERS

$\alpha_{j,k}$ = relative volatility of component j with respect to component k, defined by Eq. (5-3)

α_j = relative volatility of component j with respect to a selected reference component

ϵ = unknown molal flow rate in Example Problem 5-5

θ = correction factor for all ℓ/d values; see Eqs. (5-71) through (5-75)

μ = unknown molal flow rate in Example Problem 5-5

$\sum$ = summation on the index k to include all components

ϕ = parameter in Underwood equations for minimum reflux; see Eqs. (5-30) and (5-31)

SUBSCRIPTS

B = bottom product stream

D = distillate product stream

f = feed plate

i = iterate number in Newton iteration for θ

j = component j
k = component k
min = minimum value
m = plate m, a plate in the stripping section
N = plate N, the top plate
n = plate n, a plate in the enriching section
p = plate p, any plate in the column
R = reboiler
0 = vapor entering bottom plate
1 = plate 1, the bottom plate, or component 1, the most volatile
2 = plate 2, the second plate from the bottom, or component 2, the second most volatile

SUPERSCRIPT

0 = limiting value, as given by Eqs. (5-76) and (5-77)

REFERENCES

1. A. P. COLBURN, *Trans. A.I.Ch.E.*, **37**, 805 (1941).
2. M. R. FENSKE, *Ind. Eng. Chem.*, **24**, 482 (1932).
3. E. R. GILLILAND, *Ind. Eng. Chem.*, **32**, 1101 (1940).
4. E. R. GILLILAND, *Ind. Eng. Chem.*, **32**, 1220 (1940).
5. C. D. HOLLAND, *Multicomponent Distillation*, Englewood Cliffs, N. J., Prentice-Hall, 1963.
6. W. K. LEWIS and G. L. MATHESON, *Ind. Eng. Chem.*, **24**, 494 (1932).
7. W. N. LYSTER, S. L. SULLIVAN, Jr., D. S. BILLINGSLEY, and C. D. HOLLAND, *Petroleum Refiner*, **38**, No. 6, 221 (1959).
8. J. H. PERRY, ed., *Chemical Engineers' Handbook*, 3rd ed., New York, McGraw-Hill, 1950, p. 569.
9. C. S. ROBINSON and E. R. GILLILAND, *Elements of Fractional Distillation*, 4th ed., New York, McGraw-Hill, 1950, pp. 219ff.
10. *Ibid.*, pp. 245ff.
11. *Ibid.*, pp. 249ff.
12. *Ibid.*, pp. 265ff.
13. *Ibid.*, pp. 347ff.
14. E. W. THIELE and R. L. GEDDES, *Ind. Eng. Chem.*, **25**, 289 (1933).
15. A. J. V. UNDERWOOD, *J. Inst. Peteroleum*, **32**, 598, 614 (1946).
16. A. J. V. UNDERWOOD, *Chem. Eng. Prog.*, **44**, 603 (1948).

STUDY PROBLEMS

1. Write a computer program using the Thiele–Geddes–theta method for simulation of a multicomponent distillation column for the six-component system of Example Problem 5-6. Assume that the relative volatilities are constant as in that example problem, and assume constant molal overflow. Use the program to check the design curve given in Fig. 5-6 and the composition profiles given in Figs. 5-7 and 5-8.
2. Fit the data of Fig. 5-2 to adequate broken-line relationships between the α values and the K value for isopentane, which will represent the temperature. Modify the computer program developed in Problem 1 to include these variations in the relative volatilities. Run the modified program to determine the effect of the variations in relative volatilities on the design curve and on the composition profiles.
3. Modify the computer program developed in Problem 2 to employ the assumption of constant pseudo molal overflow and determine the effect of this change upon the computed results. Given below are the latent heats of vaporization of the hydrocarbons at 155°F, the average column temperature.

Component	*Heat of Vaporization at 155°F, Btu/lb mole*
Propane	4,460
Isobutane	6,750
n-Butane	7,640
Isopentane	9,530
n-Pentane	10,100
n-Hexane	12,750

Index